# Magnetoacoustic Polarization Phenomena in Solids

**Springer**
*New York*
*Berlin*
*Heidelberg*
*Barcelona*
*Hong Kong*
*London*
*Milan*
*Paris*
*Singapore*
*Tokyo*

V.V. Gudkov    J.D. Gavenda

# Magnetoacoustic Polarization Phenomena in Solids

With 120 Figures

 Springer

V.V. Gudkov
Institute for Metal Physics
Ural Division of the Russian Academy
  of Sciences
Yekaterinburg, 620219
Russia

J.D. Gavenda
Department of Physics
University of Texas
Austin, TX 78712
USA

Library of Congress Cataloging-in-Publication Data
Gudkov, V.V.
    Magnetoacoustic polarization phenomena in solids / V.V. Gudkov, J.D. Gavenda.
    p. cm.
    Includes bibliographical references and index.
    ISBN 0-387-95023-0 (hardcover : alk. paper)
    1. Acoustic magnetic resonance.   2. Polarization (Sound)   I. Gavenda, J.D.,
  1993–   II. Title
  QC763 .G84 2000
  538'.43—dc21                                                                00-023806

Printed on acid-free paper.

Production managed by Allan Abrams; manufacturing supervised by Jeffrey Taub.
Photocomposed copy prepared from the authors' LaTeX files.
Printed and bound by Edwards Brothers, Inc., Ann Arbor, MI.
Printed in the United States of America.

9 8 7 6 5 4 3 2 1

ISBN 0-387-95023-0 Springer-Verlag New York Berlin Heidelberg   SPIN 10750453

*To:*
*Janie Yeoman Gavenda*
*and*
*Alyosha Razletovskiy and Anya Gudkova*

# Preface

This book had its conception when Dr. Gudkov visited me in Austin in the Fall of 1995 and urged me to join him in writing a review of the field of magnetoacoustic polarization phenomena. I protested that, although my students and I had done some early work on this topic, most of the later work was done by researchers at the Institute for Metal Physics and by other investigators in the former Soviet Union. He eventually persuaded me that my initial contribution and general experience with magnetoacoustic phenomena qualified me to serve as a co-author. When I considered the fact that the extensive exploration of magnetoacoustic phenomena in the former Soviet Union was relatively unknown to Western scientists, I agreed to work with him on this project.

In order to make the material more accessible to nonspecialists, we have adopted consistent notation throughout the text and redrawn the figures from published papers in a consistent fashion.

Because our two institutions lie on opposite sides of the world, we needed some financial support to bring the book to fruition. We are very grateful to the Science Program for International Collaboration of the North Atlantic Treaty Organization for financial support through Grant No. HTECH.CRG 951549.

I have dedicated this book to my wife, Janie, in recognition of the support she has given me throughout my professional career.

J. D. Gavenda                    The University of Texas at Austin

I want to express my gratitude to all of the co-authors and colleagues who have supported me during the more than 25 years devoted to the investigation of magnetoacoustic polarization phenomena. In particular, I appreciate the collaboration of my teacher, Dr. Kirill Borisovich Vlasov, the theorist who developed the fundamentals of the phenomena; Dr. Albert Mazgarovich Burkhanov, the most talented experimentalist I ever met; and Dr. Anatoliy Bronislavovich Rinkevich, with whom the first and most difficult part of the way was covered.

V. V. Gudkov                                                 Institute for Metal Physics

# Contents

# List of the Most Important Symbols Used in This Book

| | |
|---|---|
| $\mathbf{B}$ | magnetic induction |
| $b_i$ | magnetoacoustic constants, Sec. 4.3.2 |
| $c$ | speed of light in vacuum |
| $c_{ijkl}$ | elastic coefficients (moduli), Sec. 4.1 |
| $\mathbf{D}$ | electric induction |
| $D_{mn}$ | transmission coefficients ($m, n$ designate types of transmitted and incident waves, respectively), Eq. (4.134) |
| $e$ | electron charge, Sec. 4.4.1 |
| $\mathbf{E}$ | electric field |
| $\mathcal{E}$ | electron energy, Sec. 4.4.1 |
| $\mathcal{E}_F$ | fermi energy, Sec. 4.4.1 |
| $\mathbf{e}_i$ | unit vector along the $i$ axis, Sec. 2.1 |
| $\mathbf{e}_g, \mathbf{e}_r$ | unit vectors defining the polarization of generating and receiving piezo-electric transducers, respectively, Sec. 2.1 |
| $\mathbf{F}$ | body force per unit volume, Sec. 4.1 |
| $F(q)$ | normalized nonlocal conductivity, Sec. 4.4.3 |
| $f$ | frequency in Hz; distribution function for conduction electrons, Sec. 4.4.1 |
| $\mathbf{g}_E$ | group velocity of an electromagnetic wave, Sec. 4.4.2 |
| $\mathbf{H}$ | magnetic field |
| $\mathbf{H}_a$ | effective field of anisotropy, Sec. 4.3.1 |
| $\mathbf{H}_d$ | demagnetization field, Eq. (4.22) |
| $\mathbf{H}_e$ | effective magnetic field, Eq. (4.24) |
| $\mathbf{H}_i$ | internal magnetic field, Sec. 4.3.1 |
| $H_K$ | Kjeldaas field, Eq. (4.86) |

$H_S$        saturation magnetic field, Sec. 5.3

**h**        variable magnetic field, Sec. 4.6.3

**j**        electric current density, Eq. (4.69)

$K_i$        anisotropy constants, Sec. 4.3.1

**k**        wave vector

$\mathbf{k}_D$        doppleron wave vector, Sec. 6.3

$\mathbf{k}_p^i$        wave vector of incident $p$-type elastic mode, Eq. (4.160)

$\mathbf{k}_s^r, \mathbf{k}_p^r, \mathbf{k}_l^r$        wave vectors of reflected $s$-, $p$,- and $l$-type elastic modes, respectively, Sec. 4.6.2

$\mathbf{k}_s^t, \mathbf{k}_p^t, \mathbf{k}_l^t$        wave vectors of transmitted elastic modes of quasi-$s$, -$p$, and -$l$ type, respectively, Sec. 4.6.2

$k_0$        ultrasonic wave number for $H = 0$, Sec. 4.4.3

$k^\pm$        wave number of circularly-polarized modes, Sec. 4.3.2

$k_e^\pm$        wave number of circularly-polarized elastic-like modes, Sec. 4.5

$k_D^\pm$        doppleron wave number, Sec. 6.5

$k_B$        Boltzmann constant

**l**        refraction vector, Sec. 4.6

$L$        sample length, Sec. 6.3

$l$        ultrasonic path length, Sec. 5.1;
mean free path of conduction electrons, Sec. 6.1

**M**        magnetization

$\mathbf{M}_S$        saturation magnetization, Sec. 4.3.1

$M_0$        modulus of magnetization, Sec. 4.3.1

**m**        variable magnetization (vector), Sec. 4.3.1;
effective mass (second rank tensor), Sec. 4.4.1

$m_0$        free electron mass

$m_c$        electron cyclotron mass, Eq. (4.58)

**N**        demagnetization tensor, Sec. 4.3.1

$N$        attenuation in dB, Eq. (2.27)

$N_e, N_h$        electron and hole concentrations in a metal, Sec. 4.4.2

**P**        momentum density, Sec. 4.3.2

**p**        electron quasi-momentum

$p$        parameter characterizing shear wave polarization, Eq. (2.6)

**q**        unit vector perpendicular to interfacial plane, Sec. 4.6

$q$        dimensionless parameter of nonlocality, Eq. (4.96)

$R_G$        Gaussian curvature, Sec. 6.1

$R_{mn}$        reflection coefficients ($m, n$ designate the types of reflected and incident waves, respectively), Eq. (4.134)

$S$        area of electron cyclotron orbit in **p** space, Sec. 4.4.1

$s$        phase velocity of ultrasound, Sec. 4.1

$s_0$        phase velocity of a shear ultrasonic wave at $H = 0$, Sec. 2.2

$s^\pm$        phase velocity of a circularly-polarized elastic mode, Sec. 2.2,

$s_e^\pm$        phase velocity of a circularly-polarized elastic-like mode, Sec. 4.5

$s_t, s_l$        phase velocity of transverse or longitudinal elastic waves, Sec. 4.3.2

| | |
|---|---|
| $T$ | absolute temperature |
| $T_D$ | Dingle temperature, Sec. 6.6 |
| $T_c$ | electron cyclotron period, Sec. 4.4.3 |
| $t$ | time |
| $\mathbf{U}$ | complex amplitude of elastic displacement, Sec. 4.1 |
| $\mathbf{U}_n^i$ | complex amplitude of an $n$-type incident wave, Sec. 4.6 |
| $\mathbf{U}_m^r$ | complex amplitude of an $m$-type reflected wave, Sec. 4.6 |
| $\mathbf{U}_m^t$ | complex amplitude of an $m$-type transmitted wave, Sec. 4.6 |
| $\mathbf{u}$ | elastic displacement vector, Sec. 4.1 |
| $u^\pm(z,t)$ | circular components of elastic displacement defined in the complex plane $(x, iy)$, Eq. (2.4) |
| $u_j^\pm$ | circular components of elastic displacement related to $B_j$, Sec. 2.3 |
| $V$ | amplitude of ac voltage, Sec. 2.1 |
| $V_{kj}$ | amplitude of ac voltage determined at $\psi_k$ and $B_j$, Sec. 2.3 |
| $\mathbf{v}$ | electron velocity |
| $\bar{\mathbf{v}}$ | electron velocity averaged over a cyclotron orbit, Eq. (4.68) |
| $\mathbf{v}_E$ | phase velocity of an electromagnetic wave, Sec. 4.4.2 |
| $W$ | density of Helmholtz free energy, Sec. 4.1 |
| $W_a$ | density of magnetic anisotropy energy, Eq. (4.20) |
| $W_{ex}$ | density of exchange energy, Eq. (4.16) |
| $W_L$ | density of elastic energy, Eq. (4.36) |
| $W_m$ | density of magnetic energy, Eq. (4.37) |
| $W_{me}$ | density of magnetoelastic energy, Eq. (4.32) |
| $\alpha$ | signal phase, |
| $\alpha_{kj}$ | signal phase determined at $\psi_k$ and $B_j$, Sec. 2.3 |
| $\alpha_0$ | relaxation parameter in the Gilbert equation, Eq. (4.55) |
| $\alpha_i$ | unit vector for direction of magnetization, Sec. 4.3.1 |
| $\alpha_{ijkl}$ | electroacoustic coefficients for conduction electron contribution to the elasticity of a metal, Eq. (4.74) |
| $\alpha^\pm$ | circular components of $\boldsymbol{\alpha}$, Eq. (4.80) |
| $\beta_{ijk}$ | electroacoustic coefficients: deformation conductivity, Eq. (4.74) |
| $\beta^\pm$ | circular components of deformation conductivity, Eq. (4.81) |
| $\Gamma$ | absorption coefficient, Sec. 2.1 |
| $\Gamma_0$ | absorption coefficient at $H = 0$, Sec. 2.2 |
| $\Gamma_e^\pm$ | absorption coefficient of circularly-polarized elastic modes, Sec. 2.2 |
| $\Gamma_e^\pm$ | absorption coefficient of circularly-polarized elastic-like modes, Sec. 4.5 |
| $\gamma$ | gyromagnetic ratio, Sec. 4.3.1; normalized relaxation frequency, Sec. 4.4.2 |
| $\varepsilon$ | ellipticity, Eq. (2.5) |
| $\varepsilon_{ij}$ | components of strain (deformation) tensor, Eq. (4.2) |

| | |
|---|---|
| $\eta$ | square root of the coefficient of transformation of elastic vibration energy into electromagnetic energy, Sec. 2.1; |
| | parameter of coupling between elastic and electromagnetic subsystems, defined in the local regime, Eq. (4.88) |
| $\theta$ | dimensionless time for electron motion around a cyclotron orbit, Eq. (4.58) |
| $\theta^i$, $\theta^r_m$ $\theta^t_n$ | angles of incidence, reflection, and transmission of waves designated by $m$ and $n$, Sec. 4.6.1 |
| $\kappa$ | electron wave vector |
| $\boldsymbol{\lambda}$ | deformation potential tensor, Sec. 4.4.1 |
| $\boldsymbol{\Lambda}$ | normalized deformation potential tensor, Eq. (4.63) |
| $\lambda$ | wavelength; |
| | relaxation parameter in Landau-Lifshitz equation, Eq. (4.54); |
| | one of the Lamé constants, Sec. 4.6.3 |
| $\mu$ | one of the Lamé constants, Sec. 4.6.3 |
| $\nu$ | relaxation frequency, Sec. 4.4.1 |
| $\rho$ | mass density |
| $\sigma_{ij}$ | electrical conductivity tensor component, Eq. (4.74) |
| $\sigma^{\pm}$ | circular components of electrical conductivity, Eq. (4.80), |
| $\tau_{ij}$ | tensor component of thermodynamic tension, Sec. 4.1 |
| $\tau^d_{ij}$ | tensor component of dynamic tension of electronic origin, Eq. (4.69) |
| $\tau^M_{ij}$ | tensor component of Maxwell tension, Eq. (4.172) |
| $\tau^T_{ij}$ | tensor component of total tension, Eq. (4.171) |
| $\tau^0_{ij}$ | complex amplitude of tension, Sec. 4.6.1 |
| $\tau$ | relaxation time of the conduction electron subsystem, Sec. 4.4.1 |
| $\tau_p$ | relaxation time of the elastic (phonon) subsystem, Sec. 4.3.2 |
| $\tau_s$ | relaxation time of the spin subsystem, Sec. 4.3.2 |
| $\phi$ | angle of rotation of the polarization plane (major ellipse axis), Eq. (2.5) |
| $\varphi$ | phase of elastic vibrations, Sec. 2.1 |
| $\varphi^{\pm}_j$ | phase of elastic circular vibrations related to $B_j$, Sec. 2.3 |
| $\Psi$ | angle between $\mathbf{e}_x$ and $\mathbf{e}_g$, Sec. 2.1 |
| $\psi$ | angle between $\mathbf{e}_g$ (plane of incidence) and $\mathbf{e}_r$, Sec. 2.1 |
| $\Omega$ | cyclotron frequency, Eq. (4.58) |
| $\omega$ | radian frequency in rad/s |

# 1
# Introduction

## 1.1 The Role of Physical Acoustics in the Study of Solids

It is impossible to ignore the important role played by physical acoustics in the continuing development of our understanding of the physics of solids. Convincing evidence for this statement can be found in the many articles published in numerous volumes of *Physical Acoustics: Principles and Methods* produced by W. P. Mason and succeeding editors. We begin our survey of magnetoacoustic polarization phenomena by examining some of the reasons for the pervasive role of acoustic methods in the study of solids.

Atoms that have combined to form a crystalline or amorphous solid constitute a medium capable of propagating all kinds of mechanical vibrations. The low-frequency branches of the vibrational spectra can be treated as acoustic waves propagating in a continuous medium, since they have wavelengths that are much larger than the sizes and spacings of the atoms. In addition to frequency, the characteristic parameters of these waves include wavelength (or phase velocity), amplitude, and polarization. The latter describes the trajectory of the periodic motion of a volume element, which cannot in general be represented by a constant vector. By measuring these parameters a researcher can obtain a great variety of information about the physical properties of the material under investigation.

One might think that only the mechanical properties of solids can be studied in these kinds of experiments. In fact, much more can be learned through the proper application of the methods of physical acoustics. A physical system such as a solid can be described as a set of several interacting subsystems. For example, the crystalline lattice can be considered to be a mechanical or elastic subsystem when dealing with small displacements of the volume elements from their equilibrium positions. We can ignore the magnetic moments of conduction electrons in order to treat them as a subsystem, and form a separate magnetic subsystem consisting of the magnetic moments of the electrons and nuclei, and so on.

In some cases the interactions of the subsystems are relatively weak and they can be investigated independently, providing data on the most well-known physical characteristics of solids, such as elastic moduli, electrical resistivity, and magnetic susceptibility. However, a phenomenon such as magnetostriction is already the result of the interaction between the magnetic and elastic subsystems, in spite of the fact that it is usually discussed in connection with magnetic properties. In investigating a particular phenomenon one very often must deal with two or even more of the subsystems of a solid; thus, research performed with acoustic methods can give information about much more than just the mechanical subsystem.

Modern acoustic techniques make it possible to generate mechanical vibrations over a very wide range of frequencies. They can be treated as waves propagating in a continuous medium up to a few gigahertz. At the lower end of the frequency range, the wavelengths can exceed the size of the specimen, so the vibrations can be considered to be homogeneous deformations. At the upper end, wavelengths can be of the order of 10 times the lattice parameter. The experiment can be performed so that any spatial characteristic bigger than the lattice parameters of the solid can have a value equal to the wavelength. This condition leads to strong spatial dispersion when anomalies in the physical parameters occur as a function of the wavelength.

And, of course, it goes almost without saying that acoustic techniques have one more great advantage: acoustic waves propagate in metals as well as in dielectrics, contrary to the situation for electromagnetic waves. Thus acoustic methods make it possible to obtain much more information for a wide variety of solids than simply how they respond to quasistatic mechanical forces.

## 1.2  Magnetoacoustic Polarization Phenomena in Solids

We will restrict our survey to phenomena involving small lattice displacements so that we can neglect nonlinear or anharmonic effects. For this case the phenomena are independent of the energy propagated by the wave, and its amplitude will not appear in the expressions describing them. Thus only the radian frequency $\omega$, the wavelength $\lambda$ (or, equivalently, the phase velocity s or the wave vector **k**), and the polarization are left to characterize a wave. Absorption (or attenuation) is another important parameter and can be represented by the imaginary part of either $\omega$ or **k**. We will employ the latter representation and use the symbol $\Gamma$ for the absorption coefficient.

Since solids can be described in a three-dimensional space, there can be only three independently propagating acoustic waves, known as normal modes or eigenmodes, if one considers only the elastic subsystem. The possible polarizations of the waves are determined by the symmetry of the crystal along the direction of propagation. If the symmetry elements are not changed in the experiment, the polarization will not vary, and most acoustical experiments in solids are arranged so that this situation holds; a wave of fixed frequency propagates along one of the principal crystallographic axes and only the phase velocity and absorption depend on some external parameter such as temperature, pressure, or magnetic field.

Actually, in the more general case the application of an external magnetic field can lower the symmetry and, in addition, interactions between the subsystems can result in an increase in the number of normal modes that transmit mechanical energy. These waves are called coupled modes and are characterized by both elastic vibrations and oscillations of the electromagnetic field or magnetization. Not just one, but two or more waves are generated in this case by either a piezoelectric transducer or a coil forming part of a resonant circuit placed near the specimen's surface. Superposition of the generated modes may cause the direction of vibration to change as the waves propagate. In optics such variations of the polarization depending on an external static magnetic field **H** are well known as the magneto-optical Faraday, Cotton–Mouton, and Kerr effects.

This book is devoted to reporting progress in the investigations of magnetoacoustic analogs of these phenomena. The frequencies of the waves used range from $10^7$ to $10^9$ Hz. Accordingly, we will often use the terms *ultrasonic* or *ultrasound* to indicate the frequency range of an effect under consideration.

There are two general categories of magnetoacoustic polarization phenomena (MPP). In the first, the ultrasonic wave propagates through a continuous homogeneous medium and the polarization changes are related to the bulk properties of that medium that depend on the magnetic field. In the second category, the ultrasound propagates through a nonmagnetic medium and reflects from the interface with a second medium whose properties are being explored. Changes in the polarization of the reflected wave are characteristic of the magnetic properties of the second medium.

Faraday and Cotton–Mouton effects as well as their acoustic analogs represent the most typical bulk polarization phenomena. The features that the acoustical and optical versions have in common are the following. A linearly polarized transverse wave propagates in an isotropic medium whose characteristics are independent of external magnetic fields and impinges on an interface with a second, anisotropic medium whose characteristics depend on its magnetic state. After propagating through the second medium and passing through a second interface into a third medium whose characteristics are identical to those of the first, the polarization of the wave may be changed; it may become elliptically polarized with the major ellipsoid axis inclined with respect to the polarization plane of the wave in the first medium. We shall refer to the ellipticity $\varepsilon$ (whose modulus is the ratio of the minor to the major ellipse axis) and the angle of rotation of the polarization plane, $\phi$, as the polarization parameters of the ultrasonic wave, keeping in mind that, as usual, $\phi$ is the angle between the polarization plane in the first medium and the major ellipse axis of the polarization in the third one. Thus, $\varepsilon$ and $\phi$ are quantitative characteristics of the polarization phenomena.

As a rule, the propagation direction of the ultrasound is perpendicular to the interfacial planes, and the magnetic field that determines the magnetic state of the second medium is either parallel or perpendicular to the wave vector. We refer to bulk polarization phenomena in a longitudinal magnetic field as the acoustic analog of the Faraday effect, while those in a perpendicular field we call the acoustic analog of the Cotton–Mouton effect.

The polarization parameters for MPP in a longitudinal field with ultrasonic propagation along a crystallographic axis of high-order rotational symmetry (not less than three-fold) are odd with respect to inversion of the magnetic induction $\mathbf{B}$,

$$\varepsilon(\mathbf{B}) = -\varepsilon(-\mathbf{B}), \qquad \phi(\mathbf{B}) = -\phi(-\mathbf{B}), \tag{1.1}$$

and depend on the differences in the phase velocities and absorption coefficients of the circularly polarized normal modes.

On the other hand, the polarization parameters for MPP in a perpendicular field are even with respect to $\mathbf{B}$ inversion,

$$\varepsilon(\mathbf{B}) = \varepsilon(-\mathbf{B}), \qquad \phi(\mathbf{B}) = \phi(-\mathbf{B}), \tag{1.2}$$

and are due to the differences in phase velocities and absorption coefficients of the linearly polarized normal modes.

Thus, MPP are usually described by the characteristics of the two normal modes: Those which are odd are due to circular magnetic birefringence and dichroism (the differences in phase velocities and absorption coefficients of the two circularly polarized modes, respectively), whereas those which are even are due to linear magnetic birefringence and dichroism (the differences in the same properties for linearly polarized modes).

Since materials with a low level of absorption are commonly used in optical experiments, Faraday and Cotton–Mouton effects are due to birefringence and consist in rotation of the polarization plane or ellipticity, respectively. On the other hand, dichroism as well as birefringence often plays an important role in acoustics. As a result, ultrasonic analogs of Faraday and Cotton–Mouton effects exhibit both ellipticity and rotation of the polarization plane simultaneously.

As for surface polarization phenomena, it should be noted that those at inclined or perpendicular incidence to the interface are the acoustic analogs of the magneto-optical Kerr effect.

One more circumstance should be mention to indicate more clearly the place of magnetoacoustic polarization phenomena in physics. These are effects that can be observed in solids only; transversely polarized waves cannot propagate in liquids or gases, thus no changes in wave polarization can occur.

## 1.3   A Brief Historical Overview

In 1958 Kittel [1] and Vlasov [2] independently proposed that magnetoacoustic polarization phenomena should be observable. Kittel developed a microscopic theory of the acoustic analog of the Faraday effect under magnon–phonon resonance, while Vlasov and Ishmukhamedov worked out a phenomenological theory [2–4] based on earlier papers [5, 6].

Magnon–phonon resonance is a result of the interaction of the elastic and magnetic subsystems in a magnetically ordered medium that occurs when the wave vector of the normal mode of the magnetic subsystem representing alternations of magnetization and called a *spin wave* or, in terms of quasiparticles, a *magnon*, is equal to that of the elastic wave at a particular frequency. This resonance is also known as a magnetoacoustic or magnetoelastic resonance; however, similar terms are also used with respect to other phenomena observed in nonmagnetic materials. Thus we will use the term *magnon–phonon resonance* for the interaction of magnetic and elastic subsystems in magnetically ordered materials.

In 1959 Morse and Gavenda [7] were the first to observe polarization rotation. Their experiment was performed in a copper crystal using a longitudinal magnetic field. Later Bömmel and Dransfeld [8] reported an analogous phenomenon in yttrium iron garnet (YIG). The first thorough investigation of the acoustic analog of the Faraday effect was carried out by Matthews and Le Craw [9] and published in 1962; they studied YIG under the conditions discussed by Kittel [1]. In the following year Lüthi [10] discovered the acoustic analog of the Cotton–Mouton effect in the same material. Later, the phenomena were studied in various magnetic materials [11–15].

In parallel with these experiments, various effects in a longitudinal magnetic field in pure metals have been investigated in which MPP due to conduction electrons take place at low temperatures. In 1964 Jones found rotation of the polarization of ultrasound in aluminum [16], and later pa-

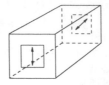

FIGURE 1.1. Positions of piezoelectric transducers on the specimen for the technique used in Ref. [20]. The polarizations of the transducers are shown by double arrows.

pers appeared with observations of this phenomenon in potassium [17, 18], copper [19–21], tin [22], aluminum [23, 24], and tungsten [25].

In 1968 the results of investigating an acoustic analog of the Faraday effect under conditions for electron spin resonance in MgO alloyed with nickel were published by Guermeur et al. [26].

The variety of systems in which polarization phenomena are observed naturally correlates with the variety of mechanisms that generate these phenomena. Of course, their most obvious manifestation takes place for any resonance: magnon–phonon [9–15], electron spin [26], and Doppler-shifted cyclotron [7, 16–25] resonances.

It should be noted that in the 1960s a number of new electromagnetic modes in metals were found. Some of them interact with ultrasound to cause helicon–phonon [27] and doppleron–phonon resonances [28–31] (HPR and DPR, respectively), thus initiating polarization effects.

However, investigations in metals are more difficult in comparison with those in magnetic dielectrics because of the large ultrasonic absorption in the former. As a result, the techniques used in the first experiments, based on the modulation of the exponential decay of the amplitudes of ultrasonic echo pulses due to rotation of the polarization [9] or to ellipticity [10], become inapplicable since only a few echo pulses can usually be observed in metal specimens.

The first step toward solving this problem was presented by Boyd and Gavenda [20]. The authors suggested a technique for deducing the angle $\phi$ under conditions where the MPP are odd under magnetic field reversal. It involved making two measurements of the amplitude of the signal that passed through the specimen, first with the magnetic field parallel to the wave vector, and then with it antiparallel. In addition, the receiving piezoelectric transducer was rotated through an angle $\psi$ with respect to the generating one (see Fig. 1.1). The applicability of this technique is limited to cases where $\varepsilon \approx 0$, which is usually valid for Doppler-shifted cyclotron resonance (DSCR).

In the meantime, theoretical papers were published by Vlasov and Filippov [32–34] in which the simultaneous existence of ellipticity and rotation of the polarization was predicted even for nonmagnetic metals. These and other papers stimulated the elaboration of a technique for measuring

$\varepsilon$, such as that developed in [35]. By adding phase measurements to the scheme used in [20], one could evaluate not only $\varepsilon$, but $\phi$ as well, even when the ellipticity is not negligible. This technique proved to have high precision and accuracies as good as 0.01 for $\varepsilon$ and better than one degree for $\phi$ were attainable. Another important feature of the method was that it made it possible for the first time to measure the magnetic field variations of phase velocities and absorption coefficients of circularly polarized ultrasonic modes.

This improved technique greatly simplified the interpretation of experimental data. For example, the interaction of ultrasound with the conduction electrons under DSCR leads to variations in velocity and absorption for both of the circularly polarized normal modes, whereas its interaction with electromagnetic modes (such as helicons and dopplerons) manifests itself in the properties of only one of the circularly polarized ultrasonic modes for a given direction of $\mathbf{H}$ with respect to $\mathbf{k}$. The first applications of the technique to the study of doppleron–phonon resonance in tungsten [36] and indium [37] gave experimental proof for the thesis that ellipticity and rotation of the polarization can occur simultaneously.

Further advances in the technique of ultrasonic polarimetry reported in [38] involve the acoustic analog of the Cotton–Mouton effect in appropriately configured experiments. That paper suggested doing the measurements with a fixed direction of $\mathbf{B}$ but with two different values of $\psi$, namely, $\psi_1 = \pi/4$ and $\psi_2 = -\pi/4$. Moreover, this method enables one to study MPP that are not characterized by the symmetry properties with respect to $\mathbf{B}$ inversion described by either of the relations (1.1) or (1.2). A generalization of this technique for arbitrary values of $\psi_1$ and $\psi_2$ was presented in [39].

Unfortunately, the method described in the latter two references requires the existence of a reference state (for a particular $\mathbf{B}$) with $\varepsilon = 0$ and $\phi = 0$. This requirement restricts its application to magneto-ordered materials. Recently a technique free of this restriction has been suggested [40].

Thus, the development of the basic techniques, the first successful experiments, and a number of theoretical papers (those previously mentioned and also [41–65]) have stimulated experimental investigations in which a number of new effects have been discovered. For example, quantum [66] and geometric [67] oscillations of $\varepsilon$ and $\phi$, pseudo-doppleron–phonon resonance [68], and circular magnetic trirefringence and trichroism of ultrasound [69] have been observed.

## 1.4  Goals of the Present Work

One might wonder whether magnetoacoustic polarization experiments provide new physical information that is inaccessible by other magnetoacoustic methods. We, of course, would respond affirmatively, since ellipticity and the angle of rotation of the polarization plane are themselves new physical parameters. If nothing else, their measured values can be useful in the design of new acoustoelectronic devices.

As for research applications, it should be pointed out that it is impossible to avoid the manifestation of polarization effects when performing certain kinds of ultrasonic experiments. This statement is particularly pertinent for shear wave propagation along a high-symmetry direction. In this case the experimentalist must deal with the rotation of polarization and the ellipticity of the ultrasonic wave when the external magnetic field is applied parallel to this axis, and must correctly take into account any polarization phenomena when analyzing measurements of the amplitude and phase of the signal.

Moreover, $\varepsilon(H)$ and $\phi(H)$ are the only parameters that can be measured in an experiment employing the Faraday effect configuration if circular magnetic trirefringence and trichroism are present. It is just these parameters that become the sources of physical information, rather than the traditional absorption or phase velocity, since the latter are impossible to determine from the experimental data in this case. The same can be said for experiments involving the Kerr effect.

In the pages that follow we describe the methods and instrumentation of ultrasonic polarimetry as well as the results of both recent and earlier experiments in order to give a rather complete survey of the variety of manifestations of magnetoacoustic polarization phenomena. This is the first review entirely devoted to this subject, although a few paragraphs about experiments in metals [70, 71] and magnetic materials [72, 73] can be found in previous review articles or books. Furthermore, there have been many interesting new developments during the twenty-odd years that have passed since those publications first appeared.

The material presented here may be divided into three parts. The first describes the methods and instrumentation used for measuring ellipticity and rotation of the polarization. We should point out that they can be applied not only to the investigation of magnetoacoustic effects, but also to any experiment in which a transverse component of the ultrasonic polarization is to be measured as a function of an external parameter such as temperature or electric field. We believe that these chapters will be useful to any experimentalist who is engaged in or who intends to establish a research program in physical acoustics.

The second part of the book deals with the theoretical description of magnetoacoustic polarization effects. Starting from the basic definitions that are required for developing the theoretical models, we derive and solve the equations that are needed to understand the papers in which new effects have been predicted or in which experimental results have been reported. Our goal has been to present the most general approach to the subject so that dielectric ferromagnets and nonmagnetic metals can be discussed in the same terms or, when that is not possible, at least to make apparent their common features. In reviewing the theoretical papers we give illustrations of the results obtained by the different authors wherever possible. The physical parameters of YIG and indium are used in model calculations for the illustrations. In addition to the previously mentioned group of prospective readers, the theoretical chapters may be helpful to students as well as to theorists who are not specialists in magnetoacoustics.

The chapters describing experimental investigations comprise the third part of this book. Here is located the most important part of our review: achievements in the investigation of magnetoacoustic polarization phenomena in solids. We summarize papers that demonstrate various manifestations of the effects, with emphasis on those that use these techniques to obtain data on the physical parameters of solids. As a consequence, engineers specializing in acoustoelectronics may find these chapters of particular interest.

This text was completed in the spring of 2000, so we could only include reviews of papers in print at that time. Unfortunately, we were not able to give full consideration to all of the related theoretical articles. With but few exceptions we have discussed the papers that are helpful in the interpretation of the experimental results. However, there are a number of magnetoacoustic polarization effects that have been predicted but still await discovery. For example, Vlasov and Filippov [44] have predicted the appearance of a longitudinal component of a wave which was initially transverse and *vice versa* in the Cotton–Mouton configuration (this effect has no analog in optics). Also Dominguez et al. [74] have directed attention to the type-II superconductors as a new class of solids for investigations of the Faraday effect. We hope that some of our readers will succeed in making new discoveries in this ever-growing field.

The writing of this book was made possible by financial support from the NATO Science Program for International Collaboration with Grant No. HTECH.CRG 951549. The authors greatly appreciate this support.

# 2

# Experimental Techniques

As mentioned in the Introduction, the first quantitative data on MPP were obtained in single crystals of YIG [9, 10]. The magnitudes of $\varepsilon$ and $\phi$ in these experiments were large enough to allow them to be determined from the distortion of the exponential decay of the ultrasonic echo pulses as observed on an oscilloscope. This method for evaluation of the polarization parameters is rather simple and will be discussed along with the description of experiments performed with it.

In this chapter we will also present more precise methods for measuring $\varepsilon$ and $\phi$. The most general technique uses apparatus that enables one to evaluate the amplitude and phase of the voltage on the receiving piezoelectric transducer. In addition, a variant of the technique that uses apparatus which only makes amplitude measurements is discussed. Both methods presuppose that steps have been taken to eliminate the effects of multiple reflections of ultrasound from the specimen's boundaries.

## 2.1 General Equations

Before we describe the various experimental techniques, it will be necessary to introduce a number of equations that relate the characteristic properties of the materials under investigation to the experimental parameters. Let us consider ultrasonic shear waves propagating along the positive direction of the $z$ axis. The elastic displacement $\mathbf{u}$ at time $t$ is written as

$$\mathbf{u}(z, t) = u_1 \mathbf{e}_1 + u_2 \mathbf{e}_2, \qquad (2.1)$$

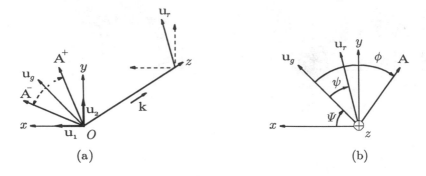

FIGURE 2.1. (a) Directions of rotation of the circular components $\mathbf{u}^{\pm}$, and the directions of the polarizations of the generating and receiving piezoelectric transducers (defined by the unit vectors $\mathbf{e}_g$ and $\mathbf{e}_r$). (b) The positive directions for defining the angles $\phi$, $\psi$, and $\Psi$.

where $\mathbf{e}_1$ and $\mathbf{e}_2$ are the unit vectors of the $x$ and $y$ axes, respectively. The vector $\mathbf{u}$ corresponds to a number $u(z,t) = u_1 + iu_2$ in the complex plane $(x, iy)$, which may be expressed as the sum of two circular components, $u^+(z,t)$ and $u^-(z,t)$. Figure 2.1 shows the directions of rotation of these components, the directions of the polarizations of the generating and receiving piezoelectric transducers (defined by the unit vectors $\mathbf{e}_g$ and $\mathbf{e}_r$, respectively), and the positive directions for defining $\phi$(the angle of rotation of the polarization), $\psi$, and $\Psi$ (the angle between $\mathbf{e}_1$ and $\mathbf{e}_g$).

Elastic vibrations excited by the generating transducer, which is located at $z = 0$, are propagated by bulk shear modes. They are assumed to have elliptical polarization (the most general case). Bear in mind that circular and linear polarizations are special cases of elliptical polarization (for which $|\varepsilon| = 1$ and $\varepsilon = 0$, respectively). Each mode, denoted by index $l$, is characterized by absorption coefficient $\Gamma_l$, ellipticity $\varepsilon_l$, wave vector $\mathbf{k}_l$, and the direction of the major ellipsoid axis, defined by the phases $\varphi_l^+$ and $\varphi_l^-$, which are independent of $z$ and $t$. The modulus of $\varepsilon_l$ is the ratio of the minor to the major ellipsoid axis, while its sign indicates the direction in which $\mathbf{e}$ moves over the elliptical trajectory: minus corresponds to movement similar to the rotation when a $(-)$-polarized mode propagates (counterclockwise if looking along the positive direction of the $z$ axis) and plus corresponds to the same for a $(+)$-polarized mode (clockwise rotation).

Elastic displacements that are due to propagation of the $l$th normal mode will be represented in the complex plane by the sum of two circular components,

$$u_l(z,t) = u_l^+(z,t) + u_l^-(z,t), \tag{2.2}$$

where

$$u_l^{\pm}(z,t) = u_l^{\pm} e^{-\Gamma_l z} \exp\left[\pm i(\omega t - k_l z \mp \varphi_l^{\pm} \pm \Psi)\right], \tag{2.3}$$

the $u_l^\pm$ are real and determine the amplitudes of the circular components, $\omega$ is the radian frequency of the ultrasound, and $k_l$ is defined by $\mathbf{k}_l = (0, 0, k_l)$.

Using these definitions we may write

$$u^\pm(z, t) = e^{\pm i\omega t} \sum_l u_l^\pm e^{-\Gamma_l z} \exp\left[\mp i(k_l z \pm \varphi_l^\pm \mp \Psi)\right]$$
$$\equiv e^{\pm i\omega t} u^\pm \exp\left[i(\Psi \mp \varphi^\pm)\right]. \tag{2.4}$$

Note that $u^+(z=0) = u^-(z=0)$ and $\varphi^+(z=0) = \varphi^-(z=0)$ because the generating piezoelectric transducer excites ultrasonic vibrations that are linearly polarized along the direction defined by $\mathbf{e}_g$.

Ellipticity and angle of rotation of the polarization are expressed in terms of $u^\pm$ and $\varphi^\pm$ in the following way:

$$\varepsilon = \frac{u^+ - u^-}{u^+ + u^-}, \qquad \phi = -\frac{1}{2}(\varphi^- - \varphi^+). \tag{2.5}$$

For our further consideration it is convenient to introduce a parameter

$$p = \frac{u^- e^{i\varphi^-}}{u^+ e^{i\varphi^+}} = (u^-/u^+) e^{i(\varphi^- - \varphi^+)}, \tag{2.6}$$

and to transform equations (2.5) into

$$\varepsilon = \frac{1 - |p|}{1 + |p|}, \qquad \phi = -\frac{1}{2} \operatorname{Im}\left[\ln(p)\right]. \tag{2.7}$$

It is clear that, in order to determine $\varepsilon$ and $\phi$, one should measure the amplitudes and the phases of the two different circular modes of ultrasonic vibrations or, to be precise, the ratio of the amplitudes and the difference in the phases of the components.

Since a piezoelectric transducer is sensitive to the vibrations that are parallel to the direction defined by $\mathbf{e}_r$, we need the projection of $\mathbf{u}(z, t)$ on $\mathbf{e}_r$:

$$u_r(z, t) = \mathbf{u} \cdot \mathbf{e}_r = \operatorname{Re}\left[(u^+ + u^-)e_r^*\right]$$
$$= \operatorname{Re}\left\{u^+ \exp\left[i(\omega t - \varphi^+ - \psi)\right] + u^- \exp\left[-i(\omega t - \varphi^- + \psi)\right]\right\}, \tag{2.8}$$

where $*$ designates the complex conjugate and $e_r = \exp\left[i(\psi + \Psi)\right]$ is the complex representation of $\mathbf{e}_r$.

Ultrasonic vibrations $u_r$ excite an ac voltage $V\cos(\omega t - \alpha) = \eta u_r$ on the receiving transducer ($\eta^2$ is the coefficient of transformation of elastic vibration energy into electric field energy, and $\alpha$ is a phase constant). Using equation (2.8) we have

$$V\cos(\omega t - \alpha) = \eta\left[u^+ \cos(\omega t - \varphi^+ - \psi) + u^- \cos(\omega t - \varphi^- + \psi)\right], \tag{2.9}$$

which may be rewritten in the form

$$\frac{V}{\eta} \left[\cos \omega t \cos \alpha + \sin \omega t \sin \alpha\right]$$

$$= \left[u^+ \cos(\varphi^+ + \psi) + u^- \cos(\varphi^- - \psi)\right] \cos \omega t$$
$$+ \left[u^+ \sin(\varphi^+ + \psi) + u^- \sin(\varphi^- - \psi)\right] \sin \omega t. \quad (2.10)$$

Since equation (2.10) is valid for arbitrary $t$, it may be transformed into two equations:

$$\frac{V}{\eta} \cos \alpha = u^+ \cos(\varphi^+ + \psi) + u^- \cos(\varphi^- - \psi), \quad (2.11)$$

$$\frac{V}{\eta} \sin \alpha = u^+ \sin(\varphi^+ + \psi) + u^- \sin(\varphi^- - \psi). \quad (2.12)$$

Multiplying equation (2.12) by $i$ and adding the result to equation (2.11) we obtain

$$\frac{V}{\eta} \exp i\alpha = u^+ \exp\left[i(\varphi^+ + \psi)\right] + u^- \exp\left[i(\varphi^- - \psi)\right]. \quad (2.13)$$

## 2.2   The Faraday Effect

The Faraday effect refers to the phenomena that obey the relations given in equations (1.1). It is observed when the direction of propagation is parallel to $\mathbf{B}$ and to an axis having $n$-fold rotational symmetry ($n \geq 3$): $\mathbf{k}_l \parallel C_n \parallel \mathbf{B} = (0, 0, B)$. The ellipticity and rotation of the polarization can be uniquely determined by making measurements for fields both parallel and antiparallel to the direction of propagation of an ultrasonic wave, as will be explained in this section.

For $B = 0$ equations (1.1) give $\varepsilon = 0$ and $\phi = 0$, in which case it follows that $u^+ = u^-$, $\varphi^+ = \varphi^-$, and equation (2.13) takes the form

$$\frac{V_0}{\eta} e^{i\alpha_0} = 2a_0 e^{i\varphi_0} \cos \psi, \quad (2.14)$$

where $a_0$ and $\varphi_0$ are the values of $u^\pm$ and $\varphi^\pm$ for $B = 0$. (Throughout this development all variables with the subscript 0 are to be evaluated at $B = 0$.)

Let us rewrite equation (2.13) for an arbitrary $B = B_1 \neq 0$, denoting $V(\mathbf{B}_1)$ as $V_1$ and $\alpha(\mathbf{B}_1)$ as $\alpha_1$:

$$\frac{V_1}{\eta} e^{i\alpha_1} = u^+(B_1) \exp\left\{i\left[\varphi^+(B_1) + \psi\right]\right\}$$
$$+ u^-(B_1) \exp\left\{i\left[\varphi^-(B_1) - \psi\right]\right\}. \quad (2.15)$$

From the Onsager relations for gyrotropic media [75] one expects

$$u^+(\mathbf{B}) = u^-(-B), \qquad \varphi^+(\mathbf{B}) = \varphi^-(-B). \tag{2.16}$$

When the direction of the magnetic field is reversed (i.e., $\mathbf{B}_2 = -\mathbf{B}_1$), we denote the transducer voltage and phase by $V_2 = V(-\mathbf{B}_1)$ and $\alpha_2 = \alpha(-\mathbf{B}_1)$ and write the equation that corresponds to (2.15) for $\mathbf{B}_2$. In accordance with equations (2.16), we replace $u_2^\pm \equiv u^\pm(-\mathbf{B}_1)$ and $\varphi_2^\pm \equiv \varphi^\pm(-\mathbf{B}_1)$ by $u_1^\mp \equiv u^\mp(\mathbf{B}_1)$ and $\varphi_1^\mp \equiv \varphi^\mp(\mathbf{B}_1)$, respectively, obtaining

$$\frac{V_2}{\eta} e^{i\alpha_2} = u_1^+ \exp\left[i\left(\varphi_1^+ - \psi\right)\right] + u_1^- \exp\left[i\left(\varphi_1^- + \psi\right)\right]. \tag{2.17}$$

After dividing the left and right sides of equations (2.15) and (2.17) by the corresponding parts of equation (2.14) we obtain the following system of equations:

$$2\cos\psi \frac{V_1}{V_0} \exp\left[i(\alpha_1 - \alpha_0)\right] = \frac{u_1^+}{a_0} \exp\left[i\left(\varphi_1^+ - \varphi_0 + \psi\right)\right]$$
$$+ \frac{u_1^-}{a_0} \exp\left[i\left(\varphi_1^- - \varphi_0 - \psi\right)\right],$$

$$2\cos\psi \frac{V_2}{V_0} \exp\left[i(\alpha_2 - \alpha_0)\right] = \frac{u_1^+}{a_0} \exp\left[i\left(\varphi_1^+ - \varphi_0 - \psi\right)\right]$$
$$+ \frac{u_1^-}{a_0} \exp\left[i\left(\varphi_1^- - \varphi_0 + \psi\right)\right]. \tag{2.18}$$

The solutions of this system of equations are

$$\frac{u_1^\pm}{a_0} \exp\left[i\left(\varphi_1^\pm - \varphi_0\right)\right]$$
$$= \frac{\frac{V_2}{V_0} \exp\left[i(\alpha_2 - \alpha_0 \mp \psi)\right] - \frac{V_1}{V_0} \exp\left[i(\alpha_1 - \alpha_0 \pm \psi)\right]}{2i\sin(\mp\psi)}. \tag{2.19}$$

Now we may determine the parameter $p$ defined in equation (2.6),

$$p = \frac{V_1 \exp\left[i(\alpha_1 - \alpha_0 - \psi)\right] - V_2 \exp\left[i(\alpha_2 - \alpha_0 + \psi)\right]}{V_2 \exp\left[i(\alpha_2 - \alpha_0 - \psi)\right] - V_1 \exp\left[i(\alpha_1 - \alpha_0 + \psi)\right]}, \tag{2.20}$$

and $\varepsilon$ and $\phi$ by using equations (2.7), once we have determined the modulus and phase angle $\varphi_p$ of the complex number $p = |p|e^{i\varphi_p}$. Note that $\operatorname{Im}(\ln p) = \varphi_p$. After some algebra we obtain

$$|p|^2 = \frac{V_1^2 + V_2^2 - 2V_1V_2\cos(\alpha_1 - \alpha_2 - 2\psi)}{V_1^2 + V_2^2 - 2V_1V_2\cos(\alpha_1 - \alpha_2 + 2\psi)} \tag{2.21}$$

and

$$\phi = \frac{1}{2} \tan^{-1} \left[ \frac{(V_1^2 - V_2^2) \sin 2\psi}{(V_1^2 + V_2^2) \cos 2\psi - 2V_1 V_2 \cos(\alpha_1 - \alpha_2)} \right]. \qquad (2.22)$$

For MPP caused by circular magnetic dichroism and birefringence it is possible to determine the magnetic-field-dependent components of the absorption $\Delta\Gamma^\pm = \Gamma^\pm(\mathbf{B}) - \Gamma_0$ and the phase velocity $\Delta s^\pm = s^\pm(\mathbf{B}) - s_0$ of the two circularly polarized normal modes:

$$\Delta\Gamma^\pm = -\frac{1}{z} \ln \frac{u^\pm}{a_0}, \qquad (2.23)$$

$$\Delta s^\pm = -\frac{s_0^2}{\omega z} (\varphi^\pm - \varphi_0). \qquad (2.24)$$

In accordance with the experimental geometry shown in Fig. 2.1, $z$ represents the distance traveled by the ultrasound as it passes through the specimen. In the chapters discussing experimental results we will use $l$ for this parameter.

According to equation (2.19),

$$\Delta\Gamma^\pm = -\frac{1}{2z} \ln \left[ \frac{V_1^2 + V_2^2 - 2V_1 V_2 \cos(\alpha_2 - \alpha_1 \mp 2\psi)}{4V_0^2 \sin^2 \psi} \right] \qquad (2.25)$$

and

$$\Delta s^\pm = -\frac{s_0^2}{\omega z} \tan^{-1} \left[ \frac{V_1 \cos(\alpha_1 - \alpha_0 \pm \psi) - V_2 \cos(\alpha_2 - \alpha_0 \mp \psi)}{V_1 \sin(\alpha_1 - \alpha_0 \pm \psi) - V_2 \sin(\alpha_2 - \alpha_0 \mp \psi)} \right]. \qquad (2.26)$$

The quantities actually measured in an experiment are

$$\Delta\alpha_i = \alpha_i - \alpha_0 \quad \text{and} \quad N_i = -20 \log \frac{V_i}{V_0}, \qquad (2.27)$$

which are the changes in the phase and the attenuation in decibels caused by the change of $B$ from 0 to $B_i$. The equations derived above can be expressed in terms of these experimental quantities by using

$$\alpha_i = \Delta\alpha_i + \alpha_0 \quad \text{and} \quad V_i = V_0 \, 10^{-N_i/20}. \qquad (2.28)$$

## 2.3   The Cotton–Mouton Effect

The second common type of MPP, the acoustic analog of the Cotton–Mouton effect, is usually observed with $\mathbf{k} \perp \mathbf{B}$. For this situation the polarization parameters have even symmetry as expressed in equation (1.2). The principal experimental techniques employ either simultaneous measurements of phase and amplitude, or amplitude alone.

### 2.3.1   The Phase-Amplitude Technique

This method consists of measuring the phase and amplitude variations of the signal for at least two different values of $B$ ($B_1$ and $B_2$) relative to an initial $B = B_0$ using three different angles for the receiving transducer: $\psi_1, \psi_2$, and $\psi_3$. Relevant equations for the three different values of $B$ and $\psi$ may be obtained by making the appropriate substitutions into equation (2.15). Here $j = 0, 1, 2$ for the three values of $B$ and $k = 1, 2, 3$ for the three values of $\psi$ for $V_{kj}$, $\alpha_{kj}$, $u_j^{\pm}$, and $\varphi_j^{\pm}$:

$$\frac{V_{10}}{\eta} e^{i\alpha_{10}} = u_0^+ \exp\left[i\left(\varphi_0^+ + \psi_1\right)\right] + u_0^- \exp\left[i\left(\varphi_0^- - \psi_1\right)\right], \quad (2.29)$$

$$\frac{V_{11}}{\eta} e^{i\alpha_{11}} = u_1^+ \exp\left[i\left(\varphi_1^+ + \psi_1\right)\right] + u_1^- \exp\left[i\left(\varphi_1^- - \psi_1\right)\right], \quad (2.30)$$

$$\frac{V_{12}}{\eta} e^{i\alpha_{12}} = u_2^+ \exp\left[i\left(\varphi_2^+ + \psi_1\right)\right] + u_2^- \exp\left[i\left(\varphi_2^- - \psi_1\right)\right]. \quad (2.31)$$

Dividing equations (2.30) and (2.31) by (2.29), we obtain

$$\frac{V_{11}}{V_{10}} \exp\left[i\left(\alpha_{11} - \alpha_{10}\right)\right] = F_1^+ e^{i\psi_1} + F_1^- e^{-i\psi_1}, \quad (2.32)$$

$$\frac{V_{12}}{V_{10}} \exp\left[i\left(\alpha_{12} - \alpha_{10}\right)\right] = F_2^+ e^{i\psi_1} + F_2^- e^{-i\psi_1}, \quad (2.33)$$

where

$$F_j^{\pm} \equiv \frac{u_j^{\pm} e^{i\varphi_j^{\pm}}}{u_0^+ \exp\left[i\left(\varphi_0^+ + \psi_1\right)\right] + u_0^- \exp\left[i\left(\varphi_0^- - \psi_1\right)\right]}. \quad (2.34)$$

Similar equations for $\psi = \psi_2$ have the form

$$\frac{V_{21}}{V_{10}} \exp\left[i\left(\alpha_{21} - \alpha_{10}\right)\right] \delta_2 \, e^{i\lambda_2} = F_1^+ \, e^{i\psi_2} + F_1^- \, e^{-i\psi_2}, \quad (2.35)$$

$$\frac{V_{22}}{V_{10}} \exp\left[i\left(\alpha_{22} - \alpha_{10}\right)\right] \delta_2 \, e^{i\lambda_2} = F_2^+ \, e^{i\psi_2} + F_2^- \, e^{-i\psi_2}, \quad (2.36)$$

where $\lambda_2$ and $\delta_2$ describe variations in phase and amplitude of the signal, respectively, caused by differences in transducer coupling to the sample while changing $\psi$ from $\psi_1$ to $\psi_2$.

One more change in $\psi$ gives the following equations in addition to (2.32)–(2.36):

$$\frac{V_{31}}{V_{10}} \exp\left[i\left(\alpha_{31} - \alpha_{10}\right)\right] \delta_3 \, e^{i\lambda_3} = F_1^+ \, e^{i\psi_3} + F_1^- \, e^{-i\psi_3}, \quad (2.37)$$

$$\frac{V_{32}}{V_{10}} \exp\left[i\left(\alpha_{32} - \alpha_{10}\right)\right] \delta_3 \, e^{i\lambda_3} = F_2^+ \, e^{i\psi_3} + F_2^- \, e^{-i\psi_3}. \quad (2.38)$$

Here $\delta_3$ and $\lambda_3$ have the same origin as $\delta_2$ and $\lambda_2$, but correspond to changing $\psi$ from $\psi_1$ to $\psi_3$.

Dividing equation (2.37) by (2.38) we obtain

$$
\begin{aligned}
F_1^+ \, e^{i\psi_3} &+ F_1^- \, e^{-i\psi_3} \\
&= \frac{V_{31}}{V_{32}} \exp\left[i(\alpha_{31} - \alpha_{32})\right] \left(F_2^+ \, e^{i\psi_3} + F_2^- \, e^{-i\psi_3}\right).
\end{aligned} \tag{2.39}
$$

Equations (2.32)–(2.33) and (2.35)–(2.36) can be solved for

$$
F_j^{\pm} = \frac{V_{1j} \exp\left(i\Delta\alpha_{1j}\right) e^{\mp i\psi_2} - L V_{2j} \exp\left(i\Delta\alpha_{2j}\right) e^{\mp i\psi_1}}{2(\mp i)V_{10}\sin(\psi_2 - \psi_1)}, \tag{2.40}
$$

where

$$
L = \delta_2 \, e^{i\lambda_2}. \tag{2.41}
$$

After substituting equation (2.40) into equation (2.39) we find

$$
L = \frac{\sin(\psi_3 - \psi_2)}{\sin(\psi_3 - \psi_1)} \frac{\frac{V_{11}}{V_{31}} \exp\left[i(\alpha_{11} - \alpha_{31})\right] - \frac{V_{12}}{V_{32}} \exp\left[i(\alpha_{12} - \alpha_{32})\right]}{\frac{V_{21}}{V_{31}} \exp\left[i(\alpha_{21} - \alpha_{31})\right] - \frac{V_{22}}{V_{32}} \exp\left[i(\alpha_{22} - \alpha_{32})\right]}, \tag{2.42}
$$

whence

$$
\begin{aligned}
|L|^2 = {}&\frac{\sin^2(\psi_3 - \psi_2)}{\sin^2(\psi_3 - \psi_1)} \\
&\times \frac{\left(\frac{V_{11}}{V_{31}}\right)^2 + \left(\frac{V_{12}}{V_{32}}\right)^2 - 2\left(\frac{V_{11}V_{12}}{V_{31}V_{32}}\right)\cos(\alpha_{11} + \alpha_{32} - \alpha_{31} - \alpha_{12})}{\left(\frac{V_{21}}{V_{31}}\right)^2 + \left(\frac{V_{22}}{V_{32}}\right)^2 - 2\left(\frac{V_{21}V_{22}}{V_{31}V_{32}}\right)\cos(\alpha_{21} + \alpha_{32} - \alpha_{31} - \alpha_{22})},
\end{aligned} \tag{2.43}
$$

$$
\begin{aligned}
\lambda_2 = {}&\tan^{-1}\left[\frac{V_{11}V_{21}}{V_{31}^2}\sin(\alpha_{11} - \alpha_{21}) - \frac{V_{11}V_{22}}{V_{32}V_{31}}\sin(\alpha_{11} + \alpha_{32} - \alpha_{22} - \alpha_{31})\right. \\
&\left. - \frac{V_{12}V_{21}}{V_{31}V_{32}}\sin(\alpha_{12} + \alpha_{31} - \alpha_{32} - \alpha_{21}) + \frac{V_{12}V_{22}}{V_{32}^2}\sin(\alpha_{12} - \alpha_{22})\right] \\
&\times \left[\frac{V_{11}V_{21}}{V_{31}^2}\cos(\alpha_{11} - \alpha_{21}) - \frac{V_{11}V_{22}}{V_{32}V_{31}}\cos(\alpha_{11} + \alpha_{32} - \alpha_{22} - \alpha_{31})\right. \\
&\left. - \frac{V_{12}V_{21}}{V_{31}V_{32}}\cos(\alpha_{12} + \alpha_{31} - \alpha_{32} - \alpha_{21}) + \frac{V_{12}V_{22}}{V_{32}^2}\cos(\alpha_{12} - \alpha_{22})\right]^{-1}.
\end{aligned} \tag{2.44}
$$

Now the parameter $p$ that appears in equation (2.7) may be obtained for an arbitrary $B = B_a$ in the form

$$
\begin{aligned}
p(B_a) &= \frac{F_a^-}{F_a^+} \\
&= \frac{V_{2a} L \exp\left[i\left(\Delta\alpha_{2a} + \psi_1\right)\right] - V_{1a} \exp\left[i\left(\Delta\alpha_{1a} + \psi_2\right)\right]}{V_{1a} \exp\left[i\left(\Delta\alpha_{1a} - \psi_2\right)\right] - V_{2a} L \exp\left[i\left(\Delta\alpha_{2a} - \psi_1\right)\right]},
\end{aligned}
$$

(2.45)

where $V_{1a}$ and $\Delta\alpha_{1a}$ are changes in the amplitude and phase of the signal due to the change of the magnetic induction from $B_0$ to $B_a$ measured for $\psi = \psi_1$; $V_{2a}$ and $\Delta\alpha_{2a}$ are those for $\psi = \psi_2$, but the changes are defined with respect to the initial state at $B = B_0$ and $\psi = \psi_1$; and $L$ is determined by equation (2.42).

Finally,

$$
|p|^2 = \frac{\delta_2^2 V_{2a}^2 + V_{1a}^2 - 2\,\delta_2\, V_{2a} V_{1a} \cos\left(\alpha_{2a} - \alpha_{1a} - \psi_2 + \psi_1 + \lambda_2\right)}{\delta_2^2 V_{2a}^2 + V_{1a}^2 - 2\,\delta_2\, V_{2a} V_{1a} \cos\left(\alpha_{2a} - \alpha_{1a} + \psi_2 - \psi_1 + \lambda_2\right)}
$$

(2.46)

and

$$
\begin{aligned}
\phi(B) = \frac{1}{2} \tan^{-1}\Big\{ &\left[\delta_2^2\, V_{2a}^2 \sin 2\psi_1 + V_{1a}^2 \sin 2\psi_2 \right. \\
&\left. - 2\,\delta_2\, V_{2a} V_{1a} \sin(\psi_1 + \psi_2) \cos\left(\alpha_{2a} - \alpha_{1a} + \lambda_2\right)\right] \\
&\times \left[\delta_2^2\, V_{2a}^2 \cos 2\psi_1 + V_{1a}^2 \cos 2\psi_2 \right. \\
&\left. - 2\delta_2 V_{2a} V_{1a} \cos(\psi_1 + \psi_2) \cos\left(\alpha_{2a} - \alpha_{1a} + \lambda_2\right)\right]^{-1} \Big\}.
\end{aligned}
$$

(2.47)

If the values of $\phi$ and $\varepsilon$ are known for a particular $B_0$ (according to our notation they are $\phi_0$ and $\varepsilon_0$), measurements at two angles $\psi_1$ and $\psi_2$ are enough to determine $L$. As a consequence of the definitions (2.5) and (2.6), for this case

$$
p_0 = \frac{1 - \varepsilon_0}{1 + \varepsilon_0}\, e^{-2i\phi_0}.
$$

(2.48)

By using equation (2.45) we find

$$
L = \frac{V_{10}\, e^{i\psi_2} + p_0\, e^{-i\psi_2}}{V_{20}\, e^{i\psi_1} + p_0\, e^{-i\psi_1}}\, e^{i(\alpha_{10} - \alpha_{20})}.
$$

(2.49)

## 2.3.2  The Amplitude Technique

In considering a variant of these experimental techniques based on amplitude measurements alone, we will restrict ourselves to discussing only the case where there is a state characterized by $\varepsilon = 0$ and $\phi = 0$ at a particular $B_0$. We begin the discussion by addressing equations (2.32), (2.35), and (2.37). After multiplying the left and right sides of these equations by their complex conjugates we obtain

$$\left(\frac{V_{11}}{V_{10}}\right)^2 = \left|F_1^+\right|^2 + \left|F_1^-\right|^2 + 2\left|F_1^+\right|\left|F_1^-\right|\cos\left(\Delta\varphi_1 + 2\psi_1\right),$$

(2.50)

$$\left(\frac{V_{21}\,\delta_2}{V_{10}}\right)^2 = \left|F_1^+\right|^2 + \left|F_1^-\right|^2 + 2\left|F_1^+\right|\left|F_1^-\right|\cos\left(\Delta\varphi_1 + 2\psi_2\right),$$

(2.51)

$$\left(\frac{V_{31}\,\delta_3}{V_{10}}\right)^2 = \left|F_1^+\right|^2 + \left|F_1^-\right|^2 + 2\left|F_1^+\right|\left|F_1^-\right|\cos\left(\Delta\varphi_1 + 2\psi_3\right),$$

(2.52)

where

$$\Delta\varphi_i = \varphi^+(B_i) - \varphi^-(B_i)$$

(2.53)

and, according to (2.49),

$$\delta_i = \frac{V_{10}\cos\left(\psi_i\right)}{V_{i0}\cos\left(\psi_1\right)}.$$

(2.54)

These operations are necessary to remove the phase $\alpha_{kj}$ from our equations since phase is not measured in this variant. We divide both sides of equations (2.50)–(2.52) by $\left|F_1^+\right|\left|F_1^-\right|$ to obtain

$$\left|p_1\right|^{-1} + \left|p_1\right| + 2\cos\left[2(\phi_1 - \psi_1)\right] = \frac{(V_{11}/V_{10})^2}{\left|F_1^+\right|\left|F_1^-\right|},$$

(2.55)

$$\left|p_1\right|^{-1} + \left|p_1\right| + 2\cos\left[2(\phi_1 - \psi_2)\right] = \frac{(V_{21}\,\delta_2/V_{10})^2}{\left|F_1^+\right|\left|F_1^-\right|},$$

(2.56)

$$\left|p_1\right|^{-1} + \left|p_1\right| + 2\cos\left[2(\phi_1 - \psi_3)\right] = \frac{(V_{31}\,\delta_3/V_{10})^2}{\left|F_1^+\right|\left|F_1^-\right|},$$

(2.57)

where $p_1 = p(B_1)$ as defined by equation (2.6).

Thus, we have three equations with three unknowns, namely, $\left|F_1^+\right|\left|F_1^-\right|$, $\left|p_1\right|$, and $\phi_1$. The latter two are the parameters we are interested in. Instead of reproducing all of the steps needed to solve the system of equations we will just present the solutions:

$$\begin{aligned}
\phi_1 = \frac{1}{2}\tan^{-1}\Big\{ &\left[(V_{21}^2\,\delta_2^2 - V_{31}^2\,\delta_3^2)\cos 2\psi_1 \right.\\
&+ \left(V_{31}^2\,\delta_3^2 - V_{11}^2\right)\cos 2\psi_2 + \left(V_{11}^2 - V_{21}^2\,\delta_2^2\right)\cos 2\psi_3\right]\\
&\times\left[(V_{21}^2\,\delta_2^2 - V_{31}^2\,\delta_3^2)\sin 2\psi_1 + \left(V_{31}^2\,\delta_3^2 - V_{11}^2\right)\sin 2\psi_2 \right.\\
&\left. + \left(V_{11}^2 - V_{21}^2\,\delta_2^2\right)\sin 2\psi_3\right]^{-1}\Big\}
\end{aligned}$$

(2.58)

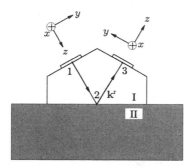

FIGURE 2.2. Schematic representation of experiments for investigating the ultrasonic analog of the Kerr effect.

and

$$|p_1| = \frac{a_1 b_1}{c_1} \pm \left[ \left( \frac{a_1 b_1}{c_1} \right)^2 - 1 \right]^{1/2}, \qquad (2.59)$$

where

$$a_1 = V_{11}^2 \sin\left[2(\psi_2 - \psi_3)\right] + V_{21}^2 \delta_2^2 \sin\left[(2(\psi_3 - \psi_1)\right]$$
$$+ V_{31}^2 \delta_3^2 \sin\left[2(\psi_1 - \psi_2)\right],$$
$$b_1 = \cos 2\phi_1,$$

$$c_1 = \left(V_{21}^2 \delta_2^2 - V_{31}^2 \delta_3^2\right) \sin 2\psi_1 + \left(V_{31}^2 \delta_3^2 - V_{11}^2\right) \sin 2\psi_2$$
$$+ \left(V_{11}^2 - V_{21}^2 \delta_2^2\right) \sin 2\psi_3.$$

The $(-)$ sign should be taken before the square root in equation (2.59), since it alone allows $|p_1| = 0$ and therefore $\varepsilon = 1$.

In summarizing, we should mention that, since the relations (1.2) were not used explicitly, the equations obtained for both of the techniques discussed in this section may be applied to determine $\varepsilon$ and $\phi$ even for phenomena that are not characterized by any symmetry properties related to inversion of $\mathbf{B}$. This means that the results of experiments performed with $\mathbf{B}$ inclined with respect to $\mathbf{k}$ by an arbitrary angle may also be interpreted with the help of these techniques.

## 2.4  The Kerr Effect

A typical arrangement for experiments measuring the Kerr effect is presented in Fig. 2.2. Medium I is isotropic as usual, while the elastic properties of medium II depend on $\mathbf{B}$. For this case a change of polarization occurs only at the interface between the media. A description of the polarization at the receiving transducer (located at 3 in the figure) may be obtained with the use of equations (2.2)–(2.7), taking into account that there are two circularly polarized normal modes and their contributions to the right side of equation (2.4) should be multiplied by $R^\pm e^{i\rho^\pm}$ after the ultrasound reflects from the interface at position 2. These coefficients describe the ellipticity and rotation of the polarization that are caused by the reflection of the ultrasound from the interface between the media. After reflecting, the positively polarized mode should become negatively polarized (defined in the system having the $z$ axis parallel to the wave vector $\mathbf{k}^r$ of the reflected wave), and vice versa for the negatively polarized mode. However, this is of no great consequence since the absorption and phase velocity are the same for these two modes in medium I. Thus,

$$
\begin{aligned}
u^\pm(z,t) &= e^{\pm i\omega t}\left(R^\pm\,e^{i\rho^\pm}\right)^{\theta(z_2-z)} u_0^\pm \exp\left[-\Gamma^\pm(z-z_1)\right]\\
&\quad \times \exp\left\{\mp i\left[k^\pm(z-z_1)\pm\varphi_0^\pm\mp\Psi\right]\right\}\\
&\equiv e^{\pm i\omega t}u^\pm\exp\left[i(\Psi\mp\varphi^\pm)\right],
\end{aligned}
\tag{2.60}
$$

where $u_0^\pm = u_0$ are the amplitudes of the modes at $z = z_1$, $\varphi_0^\pm$ are the initial phases, and $\theta(z)$ is the step function. Actually, such a transformation leads to new definitions of $u^\pm$ and $\varphi^\pm$ but does not change their meaning. As a result, the procedure presented in subsection 2.3 for finding $\varepsilon$ and $\phi$ is quite applicable and equations (2.43)–(2.49) may be used in this case also.

# 3
# Instrumentation

There are three principal types of ultrasonic installations that may be used for investigating MPP. Those based on amplitude or phase-amplitude measurements, discussed in the previous chapter, allow one to perform precise ultrasonic polarimetry, whereas another type, the ultrasonic field visualizer, gives qualitative data. The latter was developed to observe the general features of the polarization of a transverse wave with an oscilloscope. It may be employed for a preliminary analysis of the polarization in new materials to see whether MPP exist and to establish the most favorable conditions for their observation with the intent of further investigation by means of the precise methods.

Since the precise methods require quantitative data for the amplitude and, in some cases, the phase of the signal, the most suitable setups for this end usually take the form of bridge circuits because of their inherent accuracy.

The most typical experimental parameters for the installations described in this chapter are frequency range, 10 to 500 MHz; pulse duration, 0.54 to $2\,\mu s$; and specimen thickness, 14 to 10 mm.

## 3.1 Setups for Phase-Amplitude Measurements

The phase-amplitude devices applicable to ultrasonic polarimetry may be divided into two types: Installations that maintain a fixed frequency at which the ultrasonic wave is excited, and ones with a variable frequency.

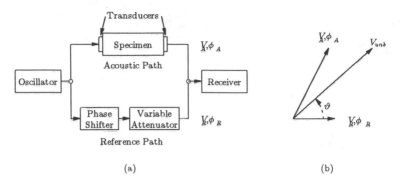

(a)                                                        (b)

FIGURE 3.1. (a) Block diagram for a fixed-frequency ultrasonic bridge. The complex amplitude and the phase of the signal from the acoustic path are $V_A$ and $\phi_A$, respectively, while those for the reference path are $V_R$ and $\phi_R$. (b) Vector diagram for the voltages at the input of the receiver. $V_{unb}$ is the complex amplitude of the error signal when the bridge is unbalanced and $\vartheta \equiv \phi_A - \phi_R$.

The first type has the advantage that the dependences of $\phi$, $\varepsilon$, $s$, or $\Gamma$ on **B** need to be measured only at a fixed frequency in a given experiment. However, since the frequency variations are quite small in the variable frequency setups, they have been extensively used owing to such advantages as simplicity of electronic design and the elimination of the calibrated phase shifter required for phase measurements with the first type.

### 3.1.1   The Fixed-Frequency Bridge

A fixed-frequency bridge has been described by Fil et al. [76]. One arm of the ultrasonic bridge circuit consists of a specimen with the piezoelectric transducers attached to it, while the other arm has a phase shifter (such as a waveguide line) and a variable attenuator [see Fig. 3.1(a)]. Both the phase shifter and the attenuator are calibrated. The signals from the two arms are combined at the input to a receiver that serves as the bridge detector. The vector sum of the acoustic and reference signals, $V_{unb} = V_A + V_R$, goes to zero as the bridge is balanced, i.e., $|V_A| = |V_R|$ and $\vartheta \equiv \phi_A - \phi_R = \pi$. These vectors are illustrated in Fig. 3.1(b).

In order to distinguish the acoustic signal following a particular path through the specimen from those following other paths and from stray electrical signals, one usually uses pulsed rather than continuous wave signals, as illustrated in Fig. 3.2. A pulse is sent to Gate A to turn on the oscillator signal for a time $t_p \approx 1.0\,\mu s$, which is short compared with the acoustic propagation time through the specimen, $T$. A second pulse is delayed by $T_d = nT$, where $n$ is an odd integer representing the number of paths taken through the sample by the desired acoustic signal. This delayed

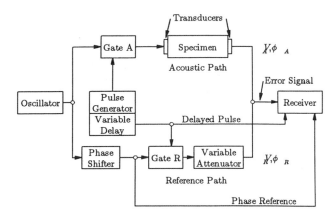

FIGURE 3.2. Modification of fixed-frequency bridge for pulsed signals.

pulse is sent to Gate R to turn on the reference signal at the time when the acoustic signal completes its path through the specimen. The delayed pulse is also used to gate appropriate signals in the receiver.

A block diagram for a receiver that is used to determine the bridge's state is shown in Fig. 3.3. The error and phase reference signals are heterodyned to a common intermediate frequency for further processing. Two phase-coherent detectors operating in quadrature allow the phase and amplitude balance conditions to be achieved independently. From Fig. 3.1(b) it can be seen that the component of $V_{unb}$ that is in phase with $V_R$ is $V_A \cos \vartheta + V_R$, while that in quadrature is $V_A \sin \vartheta$. Near balance $\cos \vartheta \approx -1$ so the output of the in-phase detector depends only on the amplitude of the error signal, while the output of the quadrature phase detector will also depend on the phase error. This allows one to distinguish between phase and amplitude unbalance caused by variation of **B** or any other external parameter. The amplitude unbalance can be canceled out by changing the loss introduced in the attenuator, while the phase unbalance is canceled out by changing the length of the waveguide line. The data obtained from the attenuator and the waveguide line are the amplitude (in logarithmic units) and phase variations of the signal, respectively. To ensure that the receiver outputs are caused by signals following a particular path through the specimen, the peak detectors that follow the phase detectors are gated by an appropriately delayed pulse.

## 3.1.2   Variable-Frequency Bridges

Variable-frequency ultrasonic bridges [77] take advantage of the fact that the wavelength of ultrasound is smaller than that of an electromagnetic wave at the same frequency by a factor of $10^5$. Hence, a small frequency

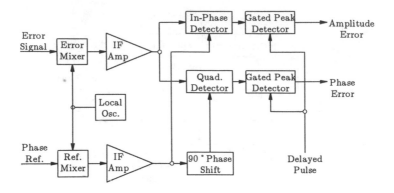

FIGURE 3.3. Block diagram for a receiver for an ultrasonic bridge.

variation leads to a rather large change in the phase of the signal that has passed through the specimen, whereas practically no phase change is seen in the other elements of the bridge circuit. The phase shifter shown in Fig. 3.2 is eliminated and phase adjustments are made by slightly varying the frequency of the oscillator. The following inequalities should be satisfied to ensure correct employment of the technique:

$$\frac{\delta\omega}{\omega} \ll \frac{\delta H}{H}, \quad \frac{\Delta s}{s} \ll 1, \quad \frac{\delta\omega}{\omega} \ll 1, \tag{3.1}$$

where $\delta\omega$ is the frequency range required to maintain the bridge balance during the experiment, $\delta H$ is the parameter describing the magnetic field variation (for instance, the line width of a certain resonance) of the parameter investigated $(s, \Gamma)$, and $\Delta s$ is the change in the ultrasonic phase velocity $s$ caused by the changes in $H$ or $\omega$ during the experiment.

The Phase-Locked Bridge

The block diagram for a device to carry out this scheme is shown in Fig. 3.4. The phase error signal from the receiver is used to automatically adjust the oscillator frequency to reduce the phase error to zero, i.e., the system forms a phase-locked loop. The amplitude error can be used to adjust the attenuator loss to reduce the amplitude error to zero.

The amplitude data are obtained from the attenuator as in the previously described device, while the phase difference $\Delta\varphi$ between two states of the bridge balance corresponding to two values of $\mathbf{B}$ ($\mathbf{B}_1$ and $\mathbf{B}_2$) may be obtained from the frequency counter using

$$\Delta\varphi = \frac{\Delta\omega}{s} z, \tag{3.2}$$

where $\Delta\omega = \omega_2 - \omega_1$, $\omega_i$ is the radian frequency when the bridge is balanced for $\mathbf{B}_i$, $z = z_0(2n-1)$ is the ultrasonic path length, $z_0$ is the sample length

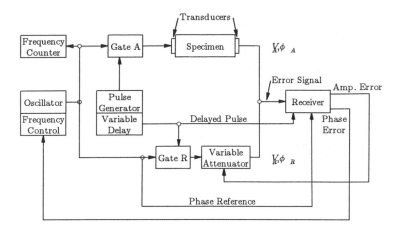

FIGURE 3.4. Block diagram for a phase-locked, variable-frequency ultrasonic bridge.

in the **k** direction, and $n$ is the number of the pulse echo chosen for the measurements.

### The Frequency-Modulated Bridge

This type of variable frequency bridge [78] was created to simplify the electronic design of the setup described above. The simplifications were achieved by introducing frequency modulation $[\omega(t) = \tilde{\omega}(1\pm m(t)/2)$, where $m(t)$ is a square wave] and using time separation of the amplitude and phase measurement processes. These modifications made it possible to omit one mixer and one intermediate-frequency (IF) amplifier, and to replace the two phase detectors operating at the intermediate frequency (about 34 MHz) with two operating at the modulation frequency (100 Hz). To accomplish this a pulse-sequence generator replaces the single pulse generator (see Fig. 3.5) and a more complex signal processing system must be employed [79].

The pulse-sequence generator controls the timing of the gate pulses and provides the square-wave signal for frequency modulation and the reference signals for the low-frequency phase detectors. Introducing low-frequency phase detectors is necessary to define the sign of the bridge unbalance because there are no RF or IF phase detectors in this scheme.

Figure 3.6 shows how the output of the amplitude detector varies as a function of frequency. When the average frequency $\tilde{\omega}$ of the oscillator is set at $\omega_0$ so that its output is being modulated between $\omega_0 - \Delta\omega/2$ and $\omega_0 + \Delta\omega/2$, the two values of the output voltage are the same ($V_0$) because of the even symmetry of the curve about the minimum corresponding to phase balance. For this case the output of the quadrature detector will be zero.

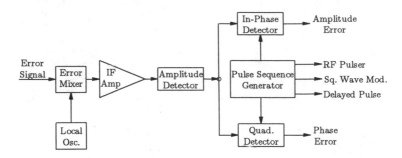

FIGURE 3.5. Block diagram for the receiver and control portion of a frequency-modulated ultrasonic bridge.

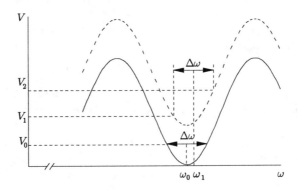

FIGURE 3.6. Voltage at the output of the amplitude detector as a function of frequency. The solid curve illustrates a phase-balanced state, while the broken curve is used to illustrate an unbalanced one. Phase balance occurs at $\omega = \omega_0$.

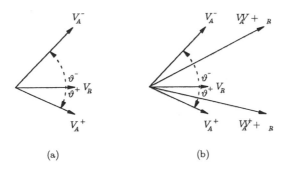

(a)                                    (b)

FIGURE 3.7. Vector diagrams of the voltages at the input of the receiver for the frequency-modulated bridge. $V_A^-$ and $V_A^+$ represent acoustic-path signals at frequencies of $\tilde{\omega}-\Delta/2$ and $\tilde{\omega}+\Delta/2$, respectively, while $V_R$ represents the reference path signal. (a) During amplitude measurements, either $|V_A^-|$ or $|V_A^+|$ is compared with $|V_R|$. (b) During phase measurements $|V_A^-+V_R|$ is compared with $|V_A^++V_R|$.

However, if $\tilde{\omega}$ is changed from $\omega_0$ to $\omega_1$, two different output voltages ($V_1$ and $V_2$) appear, associated with $\omega_1 - \Delta/2$ and $\omega_1 +\Delta/2$, respectively. The output of the quadrature phase detector is proportional to $V_1 - V_2$ and is used to adjust $\tilde{\omega}$ until phase balance is achieved. Actually, a truly balanced state never occurs because the value of the frequency that corresponds to balance ($\tilde{\omega} = \omega_0$) is the average of the modulated frequency over the period of the modulation signal.

In contrast to the previous schemes, the processes of phase and amplitude measurement are separated in time. When the in-phase detector is operational, the quadrature detector is gated off, and vice versa. In addition, the gates in the two acoustic and reference paths are used to produce different signal combinations at the input of the receiver depending on whether phase or amplitude is being measured. Some of the states that can occur are shown in Fig. 3.7.

The amplitude measurement channel operates as follows: The amplitude of the signal following the acoustic path, either $|V_A^-|$ or $|V_A^+|$, is compared by the in-phase detector with the amplitude of the signal passing through the reference path, $|V_R|$. This is accomplished by using a small number (16) of pulses corresponding to $|V_A^-|$ or $|V_A^+|$ followed by the same number of pulses corresponding to $|V_R|$ to create a square-wave signal whose amplitude indicates the amplitude unbalance of the bridge, while its phase gives the sign of this unbalance.

The phase unbalance measurements are carried out in a similar channel (the lower one in Fig. 3.5); however, the sequence of pulses as well as the way they are formed are different from those used for the amplitude measurements. These pulses are sums of the signals passed through the specimen and those passed through the attenuator that simultaneously reach the input of the receiver. The sequence consists of a small number (16)

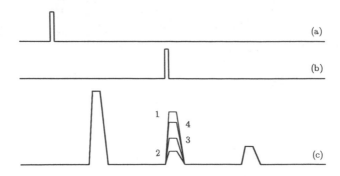

FIGURE 3.8. (a) The pulse that triggers Gate A in Fig. 3.4. (b) The pulse that triggers Gate R. (c) The output signal from the amplitude detector in Fig. 3.5. Levels 1 and 2 correspond to $|V_A^-|$ (or $|V_A^+|$) and $|V_R|$ in Fig. 3.7(a), respectively. Levels 3 and 4 correspond to $|V_A^- + V_R|$ and $|V_A^+ + V_R|$ in Fig. 3.7(b), respectively. In the amplitude-balanced state Level 1 is equal to Level 2. In the phase-balanced state Level 3 is equal to Level 4.

of pulses at the frequency $\tilde{\omega} - \Delta\omega/2$ followed by the same number of similar pulses at the frequency $\tilde{\omega} + \Delta\omega/2$ ($|V_A^- + V_R|$ or $|V_A^+ + V_R|$, respectively). After amplitude detection, a square-wave signal whose amplitude and phase contain information about the phase unbalance of the bridge is processed by the quadrature low-frequency phase detector and directed to the frequency control circuit. The latter changes the carrier frequency of the rf oscillator according to the sign of the signal received from the phase detector until $\tilde{\omega} = \omega_0$.

It is convenient to observe the state of the bridge with the aid of an oscilloscope synchronized to the pulse-sequence generator. Two pulses from this generator and the amplitude detector's output as displayed on the oscilloscope are shown in Fig. 3.8.

Some investigations, such as the acoustic analog of the Kerr effect, require the measurement of a very small variation in the amplitude of a large signal. The system just described can be modified to accomplish this by excluding the channel designed for phase measurements and adding another attenuator after the receiving transducer for more precise determination of the output level of the detector. This attenuator must be variable in small increments. The modified device will not be capable of determining phase balance but, by using sequential evaluation of signals alternately passed through one or the other arm of the bridge as described above, can make amplitude measurements.

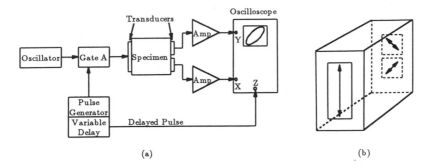

FIGURE 3.9. (a) The visualizer circuit diagram and (b) the positions of the piezo-electric transducers. The polarizations of the transducers are shown by double arrows.

## 3.2   The Ultrasonic Field Visualizer

This device was developed by Burkhanov et al. [80] to permit visual observations of the polarization of ultrasound. Its block diagram is presented in Fig. 3.9. In contrast with the usual arrangement of ultrasonic apparatus, the device has two receiving transducers. The area of each of them is one-half the area of the generating transducer. In addition, the polarization of one receiving transducer is perpendicular to the polarization of the other, while the angle between the polarization of the generating transducer and that of each receiving transducer is either $+\pi/4$ or $-\pi/4$. Two identical channels transmit the signals to the $X$ and $Y$ deflection plates of an oscilloscope whose beam is blanked except when there are acoustic signals at the $X$ and $Y$ inputs. The delayed pulse from the pulse generator is sent to the $Z$ modulation connector to accomplish this. The delay is adjusted to correspond to the ultrasonic propagation time in the specimen.

Some typical oscillograms are shown in Fig. 3.10. They were obtained while investigating the ultrasonic analog of the Cotton–Mouton effect in a nickel–ferrite specimen. The results of an investigation by means of a more precise technique are published in Ref. [81].

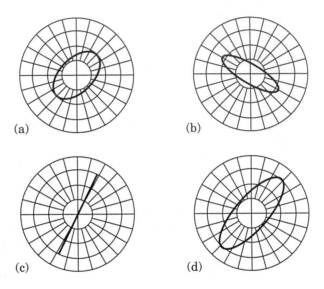

FIGURE 3.10. Polarization of the initially linear ultrasonic wave after its propagation through a specimen of $NiFe_2O_4$ as seen on the oscilloscope screen of the visualizer. The ultrasonic path length is 9.5 mm, $\omega/2\pi = 50$ MHz, and the temperature $T = 78$ K. (a) $H = 50$ Oe, (b) $H = 180$ Oe, (c) $H = 410$ Oe, and (d) $H = 520$ Oe.

# 4
# Theory

In the first three sections of this chapter we will develop a phenomenological or macroscopic theory for magnetoacoustic polarization phenomena. Model calculations for YIG are included to aid in the interpretation of the experimental results for magnetic materials presented in Chapter 5. In subsequent sections the quantum properties of the conduction electrons in metals are added to the theory to provide the basis for understanding the rich variety of MPP which have been observed in nonmagnetic metals as reported in Chapter 6. Model calculations for indium are used to illustrate how the shape of the Fermi surface determines the particular phenomena which will occur.

## 4.1   The Classical Theory of Elasticity

The classical theory of elasticity (see, for example, [82, 83] and the introductory paragraphs in [72] and [73]) contains the fundamental equations needed to describe elastic (acoustic or ultrasonic) waves in liquids and solids and to explain most of the basic phenomena. Although real objects have atomic structure, the medium in which the waves propagate is regarded as a continuum.

The principal equation of elasticity theory is the equation of motion, which relates the total force $\rho\ddot{\mathbf{u}}$ to the gradient of the stress tensor $T_{ij}$ and the body force per unit volume, $\mathbf{F}$:

$$\rho\ddot{u}_i = \frac{\partial T_{ij}}{\partial x_j} + F_i, \tag{4.1}$$

where $\rho$ is the density of the medium and $\mathbf{u}$ is the displacement of a volume element. $T_{ij}$ is defined as the force in the $\mathbf{e}_i$ direction applied to a unit square with normal vector parallel to $\mathbf{e}_j$. A similar relation holds for $T_{ij}$ replaced by the thermodynamic tensions $\tau_{ij}$, which are defined so that the sum $\tau_{ij}d\varepsilon_{ij}$ equals the differential of the work per unit of original volume done by the non-dissipative part of the stress in stretching the medium [84]. The thermodynamic tensions may be regarded as functions of the strain tensor

$$\varepsilon_{ij} = \frac{1}{2}\left(\frac{\partial u_i}{\partial x_j} + \frac{\partial u_j}{\partial x_i}\right), \tag{4.2}$$

also known as the deformation tensor, and either temperature $T$ or entropy $S$.

The ordinary elastic (or stiffness) coefficients (moduli) are defined as the first derivatives of the tensions or the second derivatives of the thermodynamic potentials with respect to strain,

$$c_{ijkl}^T = \left(\frac{\partial \tau_{ij}}{\partial \varepsilon_{kl}}\right)_T = \left(\frac{\partial^2 W}{\partial \varepsilon_{ij}\partial \varepsilon_{kl}}\right)_T \tag{4.3}$$

and

$$c_{ijkl}^S = \left(\frac{\partial \tau_{ij}}{\partial \varepsilon_{kl}}\right)_S = \left(\frac{\partial^2 U}{\partial \varepsilon_{ij}\partial \varepsilon_{kl}}\right)_S, \tag{4.4}$$

where $U(\varepsilon_{ij}, S)$ and $W(\varepsilon_{ij}, T)$ are the densities of internal energy and Helmholtz free energy, respectively. Later for simplicity we will omit the word density when we discuss the different types of energy, but bear in mind that in all of the equations energy refers to that contained in a unit volume. The index $T$ (or $S$) denotes constant temperature (or entropy). These ordinary elastic coefficients are called second order because they involve second derivatives of the energy.

The $n$th-order elastic coefficients may be defined as the $n$th partial derivatives of the energy with respect to strain. Higher-order coefficients ($n \geq 3$) should be included when the propagation of finite amplitude waves (i.e., nonlinear phenomena) is under consideration.

When investigating small-amplitude waves, one can restrict oneself to the second-order coefficients written, for example, in the adiabatic approximation. When the body force $\mathbf{F}$ is negligible, the equation of motion may be written in the form

$$\rho\ddot{u}_i = c_{ijkl}^S \frac{\partial^2 u_l}{\partial x_j \partial x_k}. \tag{4.5}$$

We will seek solutions having the form

$$u_i = U_i \exp[i(\omega t - \mathbf{k} \cdot \mathbf{r})]. \tag{4.6}$$

Substitution into equation (4.5) gives

$$\left( c_{ijkl}^S \frac{k_j k_m}{k^2} - \rho s^2 \delta_{ik} \right) u_k = 0, \tag{4.7}$$

where $k^2 = |\mathbf{k}|^2$ and $\mathbf{s}$ is the phase velocity (defined by $\mathbf{k} \cdot \mathbf{s} = \omega$) of the wave function that satisfies equation (4.7).

For nontrivial solutions the determinant of the coefficients of the system (4.7) vanishes:

$$\left| c_{ijkl}^S \frac{k_j k_i}{k^2} - \rho s^2 \delta_{ik} \right| = 0. \tag{4.8}$$

This equation determines $s$ as a function of direction. Since equation (4.8) is cubic in $s^2$, there are three velocities associated with each direction. If we introduce the symmetric second-rank tensor

$$\lambda_{ik} = c_{ijkl}^S \frac{k_j k_l}{k^2}, \tag{4.9}$$

we see that $\rho s^2$ are its eigenvalues, and the displacement vectors $\mathbf{u}$ associated with distinct eigenvalues are mutually perpendicular. Thus for each direction of $\mathbf{k}$ there can be three waves with mutually perpendicular displacement vectors and different (except in degenerate cases) phase velocities.

A degeneracy of the shear eigenmodes occurs when the waves propagate in an isotropic medium or along a high-rotational-symmetry direction of a crystal. This is the situation of interest to us, because *most magnetoacoustic polarization phenomena are observed only when an initial degeneracy of the shear normal modes is lifted by the application of an external magnetic field.* The resulting differences in phase velocities and damping lead to rotation of the polarization and ellipticity of the original linearly polarized wave.

## 4.2   Elastic, Magnetic, and Electric Variables

The great variety of elastic effects cannot be described by means of the simple stress–strain relations (4.3) – (4.4). Very often, elastic displacements of a volume element induce electric or magnetic fields. Thus other variables should be accounted for in addition to the elastic and thermal variables introduced in Section 4.1.

Remember that the electromagnetic properties of a medium are characterized by the electric field $\mathbf{E}$, electric induction $\mathbf{D}$, magnetic field $\mathbf{H}$, and magnetic induction $\mathbf{B}$. A complete description of a medium requires a set of 16 independent variables; for example, $\varepsilon_{ij}, E_k, H_k$, and $T$, whereas $\tau_{ij}, D_k, B_k$, and $S$ may be regarded as their dependent variables. Using this set of variables, the differential of $\tau_{ij}$ (or any other variable that depends on $\varepsilon_{ij}, E_k, H_k$, and $T$) can be written as

$$d\tau_{ij} = \left(\frac{\partial \tau_{ij}}{\partial \varepsilon_{kl}}\right)_{EHT} d\varepsilon_{kl} + \left(\frac{\partial \tau_{ij}}{\partial E_k}\right)_{\tau HT} dE_k$$
$$+ \left(\frac{\partial \tau_{ij}}{\partial H_k}\right)_{\tau ET} dH_k + \left(\frac{\partial \tau_{ij}}{\partial T}\right)_{\tau ET} dT. \tag{4.10}$$

The first law of thermodynamics states that the change in the internal energy of a body equals the work done by the external forces, $dW$, plus the change in heat, $dQ$: $dU = dW + dQ$. In a more-detailed form it is written as

$$dU = \tau_{ij} d\varepsilon_{ij} + E_i dD_i + H_i dB_i + T dS. \tag{4.11}$$

Using the definition of the free energy, $W = U - TS$, we have

$$dW = \tau_{ij} d\varepsilon_{ij} + E_i dD_i + H_i dB_i - S dT. \tag{4.12}$$

It is clear that $W$ is a function of $\varepsilon_{ij}$, $\mathbf{D}$, $\mathbf{B}$, and $T$, therefore

$$\tau_{ij} = \left(\frac{\partial W}{\partial \varepsilon_{ij}}\right)_{DBT}, \qquad E_i = \left(\frac{\partial W}{\partial D_i}\right)_{\varepsilon BT},$$

$$H_i = \left(\frac{\partial W}{\partial B_i}\right)_{\varepsilon DT}, \qquad S = \left(\frac{\partial W}{\partial T}\right)_{\varepsilon DB}. \tag{4.13}$$

Direct differentiation of these expressions reveals some elements of symmetry, for example:

$$\left(\frac{\partial \tau_{ij}}{\partial D_k}\right)_{BT} = \left(\frac{\partial E_k}{\partial \varepsilon_{ij}}\right)_{BT}. \tag{4.14}$$

Similar relations characterize the other coefficients. They may be expressed in matrix form as

$$\begin{pmatrix} \tau_{ij} \\ E_i \\ H_i \\ \Delta S \end{pmatrix} = \begin{pmatrix} c_{ijkl}^{DBT} & h_{ijk}^{BT} & h_{ijk}^{DT} & -\lambda_{ij}^{DB} \\ h_{ikl}^{BT} & \beta_{ik}^{\varepsilon BT} & (m^{-1})_{ik}^{\varepsilon T} & q_i^{\varepsilon B} \\ h_{ikl}^{DT} & (m^{-1})_{ik}^{\varepsilon T} & (\mu^{-1})_{ik}^{\varepsilon DT} & p_i^{\varepsilon D} \\ -\lambda_{kl}^{DB} & q_k^{\varepsilon B} & p_k^{\varepsilon D} & T^{-1}C^{\varepsilon DB} \end{pmatrix} \begin{pmatrix} \varepsilon_{kl} \\ D_k \\ B_k \\ \Delta T \end{pmatrix}. \tag{4.15}$$

Because temperature was taken to be an independent variable, the elements of the matrix that are not the derivatives of $W$ with respect to $T$ are isothermal. To obtain adiabatic coefficients one should assume $\Delta S = 0$ and then eliminate $T$.

# 4.3   Ferromagnets

In this section we will discuss the interaction of ultrasonic and spin waves in ferromagnets, i.e., the principal mechanism initiating MPP in materials of this type.

## 4.3.1   Basics of the Phenomenological Theory of Ferromagnetism

Energy of a ferromagnet

The phenomenological approach presupposes that the state of a magnetic crystal is completely described by its magnetic moment density, $\mathbf{M}(\mathbf{r})$, or magnetization, which is considered to be a spatially continuous function. Deviations from uniformity of $\mathbf{M}(\mathbf{r})$ cause increases in the crystal's energy. The principal assumption of the theory is that the only variations of $\mathbf{M}$ that can occur are those which leave its modulus $M_0$ unchanged. The macroscopic energy $W$ can be expressed in terms of spatial derivatives with respect to $\mathbf{M}$. Its main term, called the exchange energy, is

$$W_{ex}(\mathbf{r}) = A_{ijkl} \frac{\partial \alpha_i}{\partial x_k} \frac{\partial \alpha_j}{\partial x_l}, \tag{4.16}$$

where $\boldsymbol{\alpha} = \mathbf{M}(\mathbf{r})/M_0$. The form of the tensor's components depends upon the crystal's symmetry. Linear terms are absent from equation (4.16) because $W$ should be symmetric with respect to time inversion. Also, the isotropic contribution to (4.16) must be invariant under permutation of $i$ and $j$. Thus,

$$W_{ex}(\mathbf{r}) = A_{kl} \frac{\partial \alpha_i}{\partial x_k} \frac{\partial \alpha_i}{\partial x_l}. \tag{4.17}$$

For crystals of cubic symmetry, which will be discussed in more detail, $A_{kl} = A\delta_{kl}$ and

$$W_{ex}(\mathbf{r}) = A \left[ (\nabla \alpha_1)^2 + (\nabla \alpha_2)^2 + (\nabla \alpha_3)^2 \right]. \tag{4.18}$$

The total energy of a ferromagnet consists of the exchange energy, the energy of anisotropy, magnetostatic energy, and the energy of the magnetic moment in an external magnetic field $\mathbf{H}$ (Zeeman energy):

$$W(\mathbf{r}) = W_{ex} + W_a - \frac{1}{2}\mathbf{M} \cdot \mathbf{H}_d - \mathbf{M} \cdot \mathbf{H}. \tag{4.19}$$

The energy of anisotropy can be written as a power series in $\mathbf{M}$ and, moreover, as an even function of $\alpha_i$. In general, it is written in the form

$$W_a = K_{prs} \alpha_1^p \alpha_2^r \alpha_3^s = K(\mathbf{M}). \tag{4.20}$$

For a cubic crystal, the principal terms give

$$W_a = K_1 \left( \alpha_1^2 \alpha_2^2 + \alpha_2^2 \alpha_3^2 + \alpha_3^2 \alpha_1^2 \right) + K_2 \, \alpha_1^2 \alpha_2^2 \alpha_3^2. \qquad (4.21)$$

The demagnetizing field for a homogeneous $\mathbf{M}$ (which can only exist in a specimen of ellipsoidal form) is written as

$$\mathbf{H}_d = -4\pi \mathbf{N}(\mathbf{r})\mathbf{M}, \qquad (4.22)$$

where $\mathbf{N}(\mathbf{r})$ is the demagnetization tensor.

## Spin waves

Variations in magnetization are propagated by spin waves (or magnons). The classical approach to this problem is based on the equation of motion for $\mathbf{M}(\mathbf{r})$. Since invariance of its modulus $M_0$ was assumed previously, the equation of motion has the form

$$\dot{\mathbf{M}} = \gamma \mathbf{M} \times \mathbf{H}_e, \qquad (4.23)$$

where $\mathbf{H}_e$ is the effective magnetic field including the magnetization and $\gamma$ is the gyromagnetic ratio. The effective field can be considered to be the negative functional derivative of $W(\mathbf{r})$ with respect to $\mathbf{M}$. Since the energy is a function of $\mathbf{M}$ and its spatial derivatives, the effective field is

$$\mathbf{H}_e = -\frac{\partial W}{\partial \mathbf{M}} + \frac{\partial}{\partial x_i} \frac{\partial W}{\partial (\partial \mathbf{M}/\partial x_i)}. \qquad (4.24)$$

The equilibrium state for magnetization is obtained from the requirement that

$$\mathbf{M}_S \times \mathbf{H}_e = 0, \qquad (4.25)$$

where $\mathbf{M}_S$ is the saturation magnetization. Using equation (4.23) and equation (4.19) for $W$ in a cubic crystal, we find

$$\dot{\mathbf{M}} = \gamma \left[ \frac{2A}{M_0^2} \mathbf{M} \times \nabla^2 \mathbf{M} + \mathbf{M} \times \left( \mathbf{H} + \mathbf{H}_a + \mathbf{H}_d^0 + \mathbf{h} \right) \right], \qquad (4.26)$$

where $\mathbf{H}_a$ is the effective field of anisotropy, defined as $-\partial K/\partial \mathbf{M}$, $\mathbf{H}_d^0$ is the static demagnetizing field depending on the form of the sample, and $\mathbf{h}$ is the variable part of demagnetizing field (i.e., $\mathbf{H}_d = \mathbf{H}_d^0 + \mathbf{h}$). In the same way, it is convenient to define the magnetization as the sum of a static component, $\mathbf{M}_S$, and a variable one, $\mathbf{m}$, both $\mathbf{h}$ and $\mathbf{m}$ being proportional to $\exp \left[ i \left( \omega t - \mathbf{k} \cdot \mathbf{r} \right) \right]$.

At very low frequencies, $\mathbf{h}$ can be determined with the use of Maxwell's equations in the magnetostatic regime,

$$\nabla \times \mathbf{h} = 0, \qquad \nabla \cdot \mathbf{h} = -4\pi \nabla \cdot \mathbf{m}. \qquad (4.27)$$

In a small volume, located far from the specimen's boundary, the solution can be written as

$$\mathbf{h} = -\frac{4\pi}{k^2}\left(\mathbf{m} \cdot \mathbf{k}\right)\mathbf{k}. \tag{4.28}$$

For simplicity, let us neglect the transverse static demagnetizing field as well as the anisotropy. Thus, $\mathbf{M}_S$ will be parallel to the external field $\mathbf{H}$, assumed to be directed along the $z$ axis. Under these conditions,

$$\dot{\mathbf{M}} = \gamma\left[\frac{2A}{M_0^2}\mathbf{M} \times \nabla^2\mathbf{M} + \mathbf{M} \times \mathbf{H}_i - \mathbf{M} \times \frac{4\pi}{k^2}\left(\mathbf{m} \cdot \mathbf{k}\right)\mathbf{k}\right], \tag{4.29}$$

where $\mathbf{H}_i = \mathbf{H} + \mathbf{H}_a + \mathbf{H}_d^0$ is the internal static field. Assuming $\mathbf{m}$ to be small and approximately perpendicular to $\mathbf{M}_S$ ($m_3 \ll m_1 \sim m_2 \ll M_0$), we can linearize this equation and have the following expressions for $m_1$ and $m_2$:

$$m_1\left(i\omega - \frac{4\pi}{k^2}\gamma M_0 k_1 k_2\right) - \gamma m_2\left(\frac{2A}{M_0} + H_i + \frac{4\pi}{k^2}M_0 k_2^2\right) = 0,$$

$$\gamma m_1\left(\frac{2A}{M_0} + H_i + \frac{4\pi}{k^2}M_0 k_1^2\right) + m_2\left(i\omega + \frac{4\pi}{k^2}\gamma M_0 k_1 k_2\right) = 0. \tag{4.30}$$

The solution of these equations for the frequency has the form

$$\omega^2 = \gamma^2\left(\frac{2A}{M_0}k^2 + H_i\right)\left(\frac{2A}{M_0}k^2 + H_i + 4\pi M_0 \sin^2\theta_k\right), \tag{4.31}$$

where $\theta_k$ is the angle between $\mathbf{k}$ and $\mathbf{H}$. Typical curves showing the dispersion of spin waves, calculated for a YIG specimen, are given in Fig. 4.1.

Note that the result obtained is valid only for a plane spin wave, with the exception of a small region near $k = 0$, for which the boundary conditions of continuity of the magnetic potential $\Phi$ and the normal component of $\mathbf{B}$ should be verified.

## 4.3.2  Coupled Elastic and Spin Waves

The magnetoelastic interaction

Akhiezer [85] was the first to discuss the influence of the lattice vibrations on the energy of a ferromagnet. Corresponding terms originate from the dependence of the energy (such as exchange and dipole energy) on the lattice parameters which become variables when an elastic wave propagates in a crystal. Using quasiparticle notation, it can be introduced as the magnon–phonon interaction.

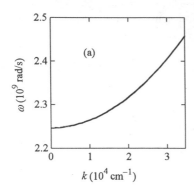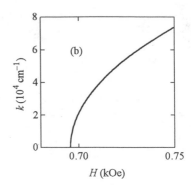

FIGURE 4.1. Dependence of (a) the frequency of spin waves on wave number at a fixed magnetic field ($H = 0.8\,\mathrm{kOe}$), and (b) the wave number on magnetic field at a fixed frequency ($\omega = 2\pi \times 6.5 \times 10^7\,\mathrm{rad/s^{-1}}$) for a spherical specimen of YIG with $\mathbf{k}\|\mathbf{H}\|[100]$.

In saturated ferromagnetic dielectrics, the energy may be expressed as a Taylor-series expansion with respect to deformations $\varepsilon_{ij}$. Under the restrictions of the linear approximation that describes magnetostriction, we can write the magnetoelastic energy as

$$W_{me} = F_{ij}(\boldsymbol{\alpha})\varepsilon_{ij} + F_{ijklmn}\frac{\partial M_n}{\partial x_i}\frac{\partial M_m}{\partial x_j}\varepsilon_{kl}. \tag{4.32}$$

Here the first term expresses the effects associated with a uniform rotation of the magnetization, while the second is determined by the inhomogeneity of $\mathbf{M}$ [86]. When investigating spin waves, the last term may be neglected as second order with respect to changes in magnetization.

Since the magnetic energy is invariant under time inversion, $F_{ij}(\boldsymbol{\alpha})$ should be an even-powered function of the components of $\boldsymbol{\alpha}$. In a cubic crystal $W_{me}$ is usually written in the form

$$
\begin{aligned}
W_{me} = {} & b_0(\varepsilon_{11} + \varepsilon_{22} + \varepsilon_{33}) \\
& + b_1\left[\left(\alpha_1^2 - \frac{1}{3}\right)\varepsilon_{11} + \left(\alpha_2^2 - \frac{1}{3}\right)\varepsilon_{22} + \left(\alpha_3^2 - \frac{1}{3}\right)\varepsilon_{33}\right] \\
& + b_2\left[\alpha_1\alpha_2\left(\varepsilon_{12} + \varepsilon_{21}\right) + \alpha_1\alpha_3\left(\varepsilon_{13} + \varepsilon_{31}\right) + \alpha_2\alpha_3\left(\varepsilon_{23} + \varepsilon_{32}\right)\right] \\
& + b_3 S\left(\varepsilon_{11} + \varepsilon_{22} + \varepsilon_{33}\right) \\
& + b_4\left[\left(\alpha_1^4 + \frac{2}{3}S - \frac{1}{3}\right)\varepsilon_{11} + \left(\alpha_2^4 + \frac{2}{3}S - \frac{1}{3}\right)\varepsilon_{22}\right. \\
& \qquad \left. + \left(\alpha_3^4 + \frac{2}{3}S - \frac{1}{3}\right)\varepsilon_{33}\right] \\
& + b_5\left[\alpha_1\alpha_2\alpha_3^2\left(\varepsilon_{12} + \varepsilon_{21}\right) + \alpha_1\alpha_3\alpha_2^2\left(\varepsilon_{13} + \varepsilon_{31}\right)\right. \\
& \qquad \left. + \alpha_2\alpha_3\alpha_1^2\left(\varepsilon_{23} + \varepsilon_{32}\right)\right] + \cdots,
\end{aligned} \tag{4.33}
$$

where $S = \alpha_1^2\alpha_2^2 + \alpha_2^2\alpha_3^2 + \alpha_1^2\alpha_3^2$ and the $b_i$ are known as the magnetoelastic constants. Sometimes an expansion of $W_{me}$ is performed with respect to the components having the symmetry of crystallographic harmonics. In this case, the coefficients are called the constants of magnetoelastic coupling and the following expression has been obtained for a cubic crystal [87]:

$$W_{me} = \beta_0 \left(\varepsilon_{11} + \varepsilon_{22} + \varepsilon_{33}\right)$$

$$+ \beta_1 \left[\left(\alpha_1^2 - \frac{1}{3}\right)\varepsilon_{11} + \left(\alpha_2^2 - \frac{1}{3}\right)\varepsilon_{22} + \left(\alpha_3^2 - \frac{1}{3}\right)\varepsilon_{33}\right]$$

$$+ \beta_2 \left(2\alpha_1\alpha_2\varepsilon_{12} + 2\alpha_2\alpha_3\varepsilon_{23} + 2\alpha_1\alpha_3\varepsilon_{13}\right)$$

$$+ \beta_3 \left(S - \frac{1}{5}\right)\left(\varepsilon_{11} + \varepsilon_{22} + \varepsilon_{33}\right)$$

$$+ \beta_4 \left[p_4(\alpha_1)\varepsilon_{11} + p_4(\alpha_2)\varepsilon_{22} + p_4(\alpha_3)\varepsilon_{33}\right]$$

$$+ \beta_5 \left[\left(7\alpha_1\alpha_2\alpha_3^2 - \alpha_1\alpha_2\right)\varepsilon_{12} + \left(7\alpha_2\alpha_3\alpha_1^2 - \alpha_2\alpha_3\right)\varepsilon_{23}\right.$$

$$\left. + \left(7\alpha_3\alpha_1\alpha_2^2 - \alpha_1\alpha_3\right)\varepsilon_{13}\right] + \cdots, \tag{4.34}$$

where $p_4(\alpha_i) = (8/35)P_4(\alpha_i) = \alpha_i^4 - (6/7)\alpha_i^2 - (6/70)$. The relations between these two sets of constants are

$$\beta_0 = b_0 + \frac{1}{5}b_3, \qquad \beta_1 = b_1 + \frac{6}{7}b_4, \qquad \beta_2 = b_2 + \frac{1}{7}b_5,$$

$$\beta_3 = b_3 + \frac{2}{3}b_4, \qquad \beta_4 = b_4, \qquad\qquad \beta_5 = \frac{2}{7}b_5. \tag{4.35}$$

Coupled magnon–phonon modes

In discussing the problem of wave propagation in a ferromagnetic crystal, one should deal with the density of the total energy $W$, which is the sum of the elastic ($W_L$), magnetic ($W_m$), and magnetoelastic ($W_{me}$) energies. It can be easily noted that the expression for $W_{me}$ consists of the products of elastic displacements and magnetization. Thus, the equations of elastic theory will contain magnetic variables, whereas the equation of motion for magnetization will contain elastic ones. These two types of equations should be combined as a system whose solutions will include coupled magnon–phonon modes, which are characterized by both elastic displacements and variations of magnetization. In fact, the magnetoelastic interaction leads to a transformation of the pure elastic waves (being the eigenmodes of the elastic subsystem) and spin waves (the eigenmodes of the magnetic subsystem) into coupled magnon–phonon waves which are the eigenmodes of the interacting subsystems. Therefore, the wave vectors of the coupled modes are functions of the parameters describing both the elastic and the magnetic properties of the crystal.

The density of the elastic energy for a cubic crystal is given in the harmonic approximation by

$$W_L = \frac{1}{2\rho} \mathbf{P(r)} \cdot \mathbf{P(r)} + \frac{1}{2} c_{1111} \left( \varepsilon_{11}^2 + \varepsilon_{22}^2 + \varepsilon_{33}^2 \right)$$
$$+ c_{1122} \left( \varepsilon_{11}\varepsilon_{22} + \varepsilon_{22}\varepsilon_{33} + \varepsilon_{11}\varepsilon_{33} \right) + 2 c_{1313} \left( \varepsilon_{12}^2 + \varepsilon_{13}^2 + \varepsilon_{23}^2 \right), \tag{4.36}$$

where $\mathbf{P(r)}$ is the momentum density. For simplicity, we will include only the isotropic exchange term and the Zeeman energy in the expression for the magnetic energy $W_m$, while the magnetoelastic energy will be represented only by terms that are of the second power in the respective $m_i$. As before, $\mathbf{H}$ is parallel to the $z$ axis. Thus,

$$W_m = \frac{A}{M_0^2} \left[ (\mathrm{grad} M_1)^2 + (\mathrm{grad} M_2)^2 + (\mathrm{grad} M_3)^2 \right] - M_3 H,$$
$$W_{me} = \frac{b_1}{M_0^2} \left( M_1^2 \varepsilon_{11} + M_2^2 \varepsilon_{22} + M_3^2 \varepsilon_{33} \right) \tag{4.37}$$
$$+ \frac{2b_2}{M_0^2} \left( M_1 M_2 \varepsilon_{12} + M_2 M_3 \varepsilon_{23} + M_1 M_3 \varepsilon_{13} \right).$$

The complete system of equations consists of the equation of elasticity theory (4.1) with $\mathbf{F} = 0$ and tensions defined as

$$\tau_{ij} = \frac{\partial (W_L + W_{me})}{\partial \varepsilon_{ij}}, \tag{4.38}$$

and the equation of motion for magnetization written in the form of equation (4.23), where $\mathbf{H}_e$ is defined by equation (4.24) with $W$ substituted for $W_m + W_{me}$.

Seeking the solution of the resulting system in the form of plane waves

$$u_i = u_i^0 \exp \left[ i \left( \omega t - \mathbf{k} \cdot \mathbf{r} \right) \right], \tag{4.39}$$

$$m_{1,2} = m_{1,2}^0 \exp \left[ i \left( \omega t - \mathbf{k} \cdot \mathbf{r} \right) \right], \tag{4.40}$$

we should linearize the system by omitting the terms containing second powers of amplitudes:

$$
\rho\omega^2 u_1 = c_{1111}k_1^2 u_1 + c_{1122}\left(k_1 k_2 u_2 + k_1 k_3 u_3\right)
$$
$$
+ c_{1313}\left(k_2^2 u_1 + k_1 k_2 u_2 + k_3^2 u_1 + k_1 k_3 u_3\right) + \frac{ib_2}{M_0}k_3 m_1,
$$
$$
\rho\omega^2 u_2 = c_{1111}k_2^2 u_2 + c_{1122}\left(k_1 k_2 u_1 + k_2 k_3 u_3\right)
$$
$$
+ c_{1313}\left(k_1^2 u_2 + k_1 k_2 u_1 + k_3^2 u_2 + k_2 k_3 u_3\right) + \frac{ib_2}{M_0}k_3 m_2,
$$
$$
\rho\omega^2 u_3 = c_{1111}k_3^2 u_3 + c_{1122}\left(k_2 k_3 u_2 + k_1 k_3 u_1\right) \tag{4.41}
$$
$$
+ c_{1313}\left(k_2^2 u_3 + k_2 k_3 u_2 + k_1^2 u_3 + k_1 k_3 u_1\right) + \frac{ib_2}{M_0}\left(k_2 m_2 + k_1 m_1\right),
$$
$$
i\omega m_1 = \gamma\left[\frac{2A}{M_0}k^2 m_2 + H\, m_2 - ib_2\left(k_3 u_2 + k_2 u_3\right)\right],
$$
$$
i\omega m_2 = \gamma\left[-\frac{2A}{M_0}k^2 m_1 - H\, m_1 + ib_2\left(k_1 u_3 + k_3 u_1\right)\right],
$$

For the special case $\mathbf{k} = (0,0,k)$ (i.e., the Faraday effect configuration) these equations reduce to

$$
\rho\omega^2 u_1 = c_{1313}k^2 u_1 + \frac{ib_2}{M_0}k m_1,
$$
$$
\rho\omega^2 u_2 = c_{1313}k^2 u_2 + \frac{ib_2}{M_0}k m_2,
$$
$$
\rho\omega^2 u_3 = c_{1111}k^2 u_3, \tag{4.42}
$$
$$
i\omega m_1 = \gamma\left(\frac{2A}{M_0}k^2 m_2 + H\, m_2 - ib_2 k u_2\right),
$$
$$
i\omega m_2 = \gamma\left(-\frac{2A}{M_0}k^2 m_1 - H\, m_1 + ib_2 k u_1\right).
$$

The use of circular components (in accordance with the definition $a^\pm \equiv a_1 \pm ia_2$) allows these equations to be written as

$$
\rho\omega^2 u^\pm = c_{1313}k^2 u^\pm + \frac{ib_2}{M_0}k m^\pm, \tag{4.43}
$$
$$
\rho\omega^2 u_3 = c_{1111}k^2 u_3, \tag{4.44}
$$
$$
\omega m^\pm = \gamma\left(\pm\frac{2A}{M_0}k^2 m^\pm \mp H\, m^\pm \pm ib_2 k u^\pm\right). \tag{4.45}
$$

Setting the determinant of the system equal to the zero yields a nontrivial solution for the variables $m^\pm$, $u^\pm$, and $u_3$. In addition, this requirement gives the dispersion equation (i.e., $\omega$ as a function of $\mathbf{k}$) for the waves that can propagate in the medium being discussed. In our case there are three dispersion equations corresponding to the three types of eigenmodes:

$$\omega^2 - s_l k^2 = 0, \tag{4.46}$$

$$\left(\omega^2 - s_t k^2\right)\left(\omega + \omega_s\right) + \frac{|\gamma| b_2^2 k^2}{\rho M_0} = 0, \tag{4.47}$$

$$\left(\omega^2 - s_t k^2\right)\left(\omega - \omega_s\right) - \frac{|\gamma| b_2^2 k^2}{\rho M_0} = 0, \tag{4.48}$$

where $s_l^2 = c_{1111}/\rho$, $s_t^2 = c_{1313}/\rho$ are the squares of the phase velocities of longitudinal and transverse ultrasonic waves, respectively, in the absence of the magnetoelastic interaction, and $\omega_s = |\gamma| H + 2A |\gamma| k^2 / M_0 \equiv \omega_0 + Y k^2$; $\gamma$ was assumed to be negative.

Equation (4.46) originates from equation (4.44) and describes a pure longitudinal elastic wave. Equations (4.47) and (4.48), from equations (4.43) and (4.45), are dispersion equations of transverse left- ($-$ index) and right- ($+$ index) circularly polarized waves. The vectors $\mathbf{m}^+$ and $\mathbf{u}^+$ rotate clockwise, while $\mathbf{m}^-$ and $\mathbf{u}^-$ rotate counterclockwise looking in the direction of wave propagation.

Assuming $b_2 = 0$, one can see that equation (4.47) has one positive solution corresponding to a pure ($-$)-polarized elastic mode, while equation (4.48) has two: one associated with a ($+$)-polarized elastic wave and a second with a spin wave of the same polarization. A finite but small value of $b_2$ does not change the solution of equation (4.47) seriously. At the same time, solutions of equation (4.48) undergo noticeable changes, especially in the region of their crossover. The crossover disappears and the dispersion curve of the former elastic wave transforms into that of a spin wave, and vice versa. Far from the crossover the curves preserve their original form, remaining quasi-elastic (or elastic-like) and quasi-spin (or spin-, magnon-like) waves. On the other hand, in the region of their strong interaction, where the largest changes are observed, the waves become coupled magnon–phonon modes that transport both elastic and magnetic forms of energy.

When polarization phenomena are under consideration, it is more convenient to solve for $\mathbf{k}(\omega)$ instead of $\omega(\mathbf{k})$. With this assumption, equations (4.47) and (4.48) can be rewritten as

$$(k^{\pm})^4 - (k^{\pm})^2 \left(\frac{\omega^2}{s_t^2} - \frac{\omega_0 \mp \omega}{Y} + \frac{\gamma b_2^2}{Y M_0 \rho s_t^2}\right) - \frac{\omega^2 (\omega_0 \mp \omega)}{Y s_t^2} = 0. \tag{4.49}$$

Solutions of this equation near resonance describe the dispersion of coupled magnon–phonon modes. The resonance, as shown in Fig. 4.2, causes a gap to appear in the dispersion curves of spin and elastic waves.

Now let us discuss the case where there is an arbitrary angle between $\mathbf{H}$ and $\mathbf{k}$, the direction of wave propagation. In addition, we will take the demagnetizing field into account. This can be done by substituting the static field $H$ for $H_i$ and inserting the linear terms of the product $\mathbf{M} \times \mathbf{h}$ into

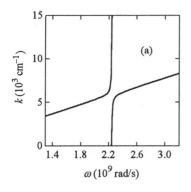
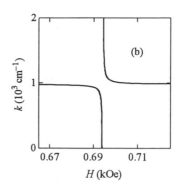

FIGURE 4.2. (a) The wave numbers of the coupled magnon–phonon modes of (+) polarization as functions of frequency at fixed magnetic field ($H = 0.8$ kOe), and (b) as functions of magnetic field at fixed frequency ($\omega = 2\pi \times 6 \times 10^7$ sec$^{-1}$), obtained for a spherical specimen of YIG with $\mathbf{k}\|\mathbf{H}\|[100]$.

the last two equations of the system (4.41). In addition, we will consider a crystal with isotropic elastic properties. It will have only two independent elastic moduli, which are related by $c_{1111} - c_{1122} = c_{1313}$. There will be two pure degenerate transverse modes and one pure longitudinal mode propagating in any direction. The polar angle (the angle between $\mathbf{k}$ and the $z$ axis) will be designated $\theta$, and $\mathbf{k}$ will be chosen to lie in the $x$-$z$ plane so that $k_x = k\sin\theta$, $k_y = 0$, and $k_z = k\cos\theta$. After applying these substitutions and definitions, we have

$$\rho\omega^2 u_1 = c_{1111}k^2\left(u_1\sin^2\theta + u_3\sin\theta\cos\theta\right)$$
$$+ c_{1313}k^2\left(u_1\cos^2\theta - u_3\sin\theta\cos\theta\right) + \frac{ib_2^2}{M_0}km_1\cos\theta,$$
$$\rho\omega^2 u_2 = c_{1313}k^2 u_2 + \frac{ib_2^2}{M_0}km_2\cos\theta,$$
$$\rho\omega^2 u_3 = c_{1111}k^2\left(u_3\cos^2\theta + u_3\sin\theta\cos\theta\right)$$
$$+ c_{1313}k^2\left(u_3\sin^2\theta - u_1\sin\theta\cos\theta\right) + \frac{ib_2^2}{M_0}km_1\sin\theta,$$
$$i\omega m_1 = \gamma\left(\frac{2A}{M_0}k^2 m_y + H_i m_2 - ib_2 k u_2\cos\theta\right), \qquad (4.50)$$
$$i\omega m_2 = \gamma\left[-\frac{2A}{M_0}k^2 m_1 - H_i m_1 - 4\pi M_0 m_1\sin^2\theta\right.$$
$$\left. + b_2 k\left(u_3\sin\theta + u_1\cos\theta\right)\right].$$

Now it will be useful to replace the elastic variables $u_1$ and $u_3$ by $u_l = u_3\cos\theta + u_1\sin\theta$ and $u_t = -u_3\sin\theta + u_1\cos\theta$, which are longitudinal and transverse displacements with respect to $\mathbf{k}$. After this, setting the determinant of the resulting system equal to zero gives us the dispersion equation of the eigenmodes:

$$\left(\rho\omega^2 - c_{1111}k^2\right)\left[\left(\rho\omega^2 - c_{1313}k^2\right)\left(\omega^2 - \omega_s\omega_m\right)\right.$$
$$- \left(\rho m^2 - c_{1313}k^2\right)|\gamma|\frac{b_2^2}{M_0}k^2\left(\omega_m \cos^2\theta + \omega_s \cos^2 2\theta\right)$$
$$\left. - |\gamma|^2\frac{b_2^4}{M_0^2}k^4\cos^2\theta\cos^2 2\theta\right]$$
$$- \left(\rho\omega^2 - c_{1313}k^2\right)\frac{b_2^2}{M_0}k^2|\gamma|\left[\omega_s\left(\rho\omega^2 - c_{1313}k^2\right)\sin^2 2\theta\right.$$
$$\left. + |\gamma|\frac{b_2^2}{M_0^2}k^2\cos^2\theta\sin^2 2\theta\right] = 0, \qquad (4.51)$$

where

$$\omega_m = |\gamma|\left(\frac{2Ak^2}{M_0} + H_i + 4\pi M_0\sin^2\theta\right), \quad \omega_s = |\gamma|\left(\frac{2Ak^2}{M_0} + H_i\right).$$

It should be noted that the longitudinal elastic mode is also coupled with the spin wave, except in the two special cases $\theta = 0$ and $\theta = \pi/2$ (i.e., $\mathbf{k} \parallel \mathbf{H}$ or $\mathbf{k} \perp \mathbf{H}$).

As for transverse waves in the case of interest, $\mathbf{k} \perp \mathbf{H}$, which corresponds to the conditions for the Cotton–Mouton effect, we have

$$\left(\rho\omega^2 - c_{1111}k^2\right)\left(\rho\omega^2 - c_{1313}k^2\right)$$
$$\times\left[\left(\rho\omega^2 - c_{1313}k^2\right)\left(\omega^2 - \omega_s\omega_m\right) - |\gamma|\frac{b_2^2}{M_0}k^2\omega_s\right] = 0. \quad (4.52)$$

For $\theta = \pi/2$ we find that $u_2$, like $u_l$, is not coupled with the spin wave, while the dispersion of the transverse coupled modes is described by the following equation [deduced from the term in square brackets in equation (4.52)]:

$$\omega^2 = \frac{1}{2}\left\{v_t^2 k^2 + \omega_s\omega_m \pm \left[\left(v_t^2 k^2 + \omega_s\omega_m\right)^2 + \frac{4|\gamma|}{\rho M_0}b_2^2 k^2\omega_s\right]^{\frac{1}{2}}\right\}. \quad (4.53)$$

When $b_2 = 0$, this expression corresponds to dispersion of pure spin and pure elastic waves, both polarized along the $x$ axis.

Until now we have been interested only in the dispersion of the coupled modes, neglecting their absorption. This was done because the main purpose was to show the approach to a description of the interaction between the magnetic and elastic subsystems of a medium, and the reasons for the existence and the main features of coupled waves. In practice, both the dispersion and the absorption of the eigenmodes produce manifestations of MPP, thus we will make a few comments about how losses can be included in the scheme presented above.

A phenomenological approach is to introduce a finite relaxation time for the spin, $\tau_s$, as well as for the elastic (or phonon) subsystem, $\tau_p$. To account for the energy losses in the magnetic subsystem one must add a relaxation term $\mathbf{R}$ to the equation of motion (4.23). When it is expressed in the form $\mathbf{R} = -\lambda \mathbf{M} \times (\mathbf{M} \times \mathbf{H}_e)$ [88], the resulting equation of motion

$$\dot{\mathbf{M}} = \gamma \mathbf{M} \times \mathbf{H}_e - \lambda \mathbf{M} \times (\mathbf{M} \times \mathbf{H}_e) \qquad (4.54)$$

is called the Landau–Lifshitz equation. It can only be used when the energy loss is small; it cannot be used to investigate the relaxation-caused shift of homogeneous ferromagnetic resonance or distortion of the spin wave dispersion curve.

Gilbert has avoided this restriction by introducing a relaxation term analogous to viscous damping [89],

$$\dot{\mathbf{M}} = \gamma \mathbf{M} \times \mathbf{H}_e + \frac{\alpha_0}{M_0} \mathbf{M} \times \dot{\mathbf{M}}, \qquad (4.55)$$

where $\alpha_0 = 1/\omega^* \tau_s = \Delta H / H_f$ is the relaxation parameter, $\omega^*$ is a parameter with units of radian frequency, $\tau_s$ is the relaxation time of the spin system, $\Delta H$ is the half-width of the absorption resonance, and $H_f$ is the field at which homogeneous ferromagnetic resonance occurs.

There is one more scheme for taking energy dissipation in the magnetic subsystem into account. It was suggested by Bloembergen and Wang [90]. A term corresponding to $\mathbf{R}$ contains two parameters characterizing the relaxation process. We will not discuss this approach since it treats the size of the magnetization as a variable and thus leads to a form of the equation of motion applicable to nonlinear phenomena. Remember that we are restricting our considerations to linear magnetization theory, wherein we assume a constant value of $M_0$. Both equations (4.54) and (4.55) match this criterion.

When magnetic metals are under consideration, one should take into account an additional reason for absorption and dispersion of coupled waves, namely, electric currents caused by the precession of the magnetization and shifts of the domain walls. Of course, a more complicated system requires a more complex description, although the principal features of coupled magnon–phonon modes will be the same. However, large damping of the waves results in essential changes in the character of the dispersion curves. Such a case will be discussed later, in Section 4.4.3.

## 4.4   Nonmagnetic Metals

In this section we will investigate pure metals that do not have magnetic ordering. In such materials, the subsystem formed by the conduction electrons needs to be accounted for in the same way as the elastic subsystem, since the interaction of elastic waves with the conduction electrons and, in particular, the electromagnetic waves that accompany the electric currents, leads to a large variety of magnetoacoustic effects, including MPP. All of these effects require a rather large relaxation time for the electronic subsystem, which can be reached at low temperatures.

Thus, we will start with a description of the electron subsystem and the electromagnetic eigenmodes in metals. Keeping in mind that, in the future, we intend to consider their interaction with elastic waves, we will restrict ourselves to frequencies of not more than $10^9$ Hz, or the gigahertz range.

### 4.4.1   Fundamentals of the Electron Theory of Metals

The modern view of a metal describes it as a lattice of ions containing a system of conduction electrons capable of traveling throughout the specimen. The character of the electron motion is conditioned by the ionic electric potential, which reproduces the periodicity of the lattice. The influence of the potential is accounted for by the effective mass $\mathbf{m}$, a second rank tensor, which is introduced into the expression for the electron energy $\mathcal{E}$. In addition, $\mathbf{m}$ defines the quasi-momentum of an electron, $\mathbf{p} = \mathbf{m} \cdot \mathbf{v}$, where $\mathbf{v} = \partial\mathcal{E}/\partial\mathbf{p}$ is the electron velocity. The ensemble of conduction electrons obeys the Pauli principle and is characterized by Fermi–Dirac statistics. As a result, at $T = 0\,\mathrm{K}$ all of the energy states in the interval $0 \leq \mathcal{E} \leq \mathcal{E}_F$ are occupied. $\mathcal{E}_F$ is called the Fermi energy, and the equation $\mathcal{E}(\mathbf{p}) = \mathcal{E}_F$ defines the Fermi surface (FS) in $\mathbf{p}$-space, which is one of the principal characteristics of a metal.

The initially sharp boundary between the empty and occupied states is broadened at higher temperatures, and electron transitions between energies near $\mathcal{E}_F$ are responsible for all of the properties peculiar to metals (see, for example, [91]).

It is obvious that magnetoacoustic effects will manifest themselves more clearly when an electron manages to complete its periodic motion in an external magnetic field. This occurs in an ultrapure perfect crystal at rather low temperatures when the relaxation time $\tau$ of the distribution function and the electron mean free path $\ell$ are large enough.

A complete system describing electromagnetic, elastic, and electron variables in metals consists of Maxwell's equations, the equations of elasticity theory, and the kinetic equation for the distribution function of conduction electrons. This system of equations has been discussed by many authors (see, for example, [42, 44, 46–49, 92–96]). The distribution function $f$ is the solution of Boltzmann's equation written in the relaxation time approximation,

$$\frac{\partial f}{\partial t} + \mathbf{v} \cdot \frac{\partial f}{\partial \mathbf{r}} + \frac{dp_z}{dt} \frac{\partial f}{\partial p_z} + \Omega \frac{\partial f}{\partial \Theta} + \frac{d\mathcal{E}}{dt} \frac{\partial f}{\partial \mathcal{E}} = -\frac{f - f_0}{\tau},$$

(4.56)

where

$$f_0 = \frac{1}{e^{[\mathcal{E}(\mathbf{p}) - \mathcal{E}_F]/k_B T]} + 1}$$

(4.57)

is the equilibrium distribution function, $k_b$ is Boltzmann's constant, and $\tau$ is the average relaxation time. The dimensionless time of electron motion over the cyclotron trajectory, $\Theta$, the cyclotron frequency $\Omega$, and the cyclotron mass $m_c$ of the electron are defined by

$$\Theta = \Omega t, \quad \Omega = -\frac{eB}{m_c c}, \quad m_c = \frac{1}{2\pi} \frac{\partial S}{\partial \mathcal{E}},$$

(4.58)

where $e < 0$ is the electron charge, $S$ is the area of an electron cyclotron orbit in $\mathbf{p}$ space, and $c$ is the velocity of light in vacuum. The $z$ axis, as before, is directed along the external magnetic field $\mathbf{H}$, which in nonmagnetic metals is, in fact, equal to the magnetic induction $\mathbf{B}$.

If we consider only harmonic changes of the variables and represent the distribution function by a sum of nonequilibrium (proportional to $\exp[i(\omega t - \mathbf{k} \cdot \mathbf{r})]$) and equilibrium components, we have

$$f(\mathbf{r}, \mathbf{p}, t) = f_0 - \chi(\mathbf{r}, \mathbf{p}, t) \frac{\partial f_0}{\partial \mathcal{E}}.$$

(4.59)

In the linear approximation we should neglect the terms proportional to $dp_z/dt$ and $d\mathcal{E}/dt$ as being of second order with respect to the variables. Thus equation (4.56) will have the form

$$\frac{\partial \chi}{\partial t} + \mathbf{v} \cdot \frac{\partial \chi}{\partial \mathbf{r}} + \Omega \frac{\partial \chi}{\partial \Theta} + \nu \chi = Q,$$

(4.60)

where $\nu = \tau^{-1}$,

$$Q = \Lambda_{ij} \dot{\varepsilon}_{ij} + e v_i F_i,$$

(4.61)

$$\mathbf{F} = \mathbf{E} + \frac{1}{c} \dot{\mathbf{u}} \times \mathbf{H} - \frac{m \ddot{\mathbf{u}}}{e}$$

(4.62)

is the effective electric field, and

$$\Lambda = \lambda - \bar{\lambda}$$

(4.63)

is the normalized deformation potential tensor, the average of which over the FS vanishes. The deformation potential $\lambda(\mathbf{p})$, introduced by Akhiezer

[97], takes into account the variation of the electron's energy in a distorted lattice (caused, in our case, by elastic wave propagation). It is defined in a coordinate system associated with the lattice (i.e., the system that moves and changes scale in accordance with the deformation of the lattice). The necessity of introducing the lattice system in addition to that of the laboratory is the result of the fact that the dispersion law for electrons $[\mathcal{E} = \mathcal{E}(\mathbf{r}, \mathbf{p}, t)]$ can only be written conveniently in such a system (variables in the lattice system are marked by a prime):

$$\mathcal{E}'(\mathbf{r}', \mathbf{p}', t) = \mathcal{E}_0(\mathbf{p}') + \lambda_{ij}(\mathbf{p}') \frac{\partial u_i}{\partial x_j} - \mathbf{v}' \cdot \frac{e}{c}(\mathbf{A} - \mathbf{u} \times \mathbf{B}) - m_0 \dot{\mathbf{u}} \cdot \mathbf{v}', \tag{4.64}$$

where $\mathcal{E}_0$ is the energy of an electron for zero deformation, $m_0$ is the free electron mass, and $\mathbf{A}$ is the vector potential of the electromagnetic field accompanying the elastic wave. The last term, which comes from the time dependence of the new coordinate system, is quite small compared with the others and may be neglected.

The solution of Boltzmann's kinetic equation may be written as

$$\chi = \tilde{Q}(\Theta), \tag{4.65}$$

where

$$\tilde{Q}(\Theta) = \int_0^{2\pi} Q(\Theta_1) G(\Theta, \Theta_1) \, d\Theta_1 \tag{4.66}$$

and $G$ is the Green function for equation (4.60):

$$G(\Theta, \Theta_1) = \frac{1}{2\pi} \left\{ \exp\left[-\frac{i}{\Omega} \int_{\Theta_1}^{\Theta} \mathbf{k}\left[\mathbf{v}(\Theta_2) - \bar{\mathbf{v}}\right] d\Theta_2 \right] \right\} \times \sum_{n=-\infty}^{\infty} \frac{e^{-in(\Theta - \Theta_1)}}{\nu - i\left(\mathbf{k} \cdot \bar{\mathbf{v}} - \omega - n\Omega\right)}. \tag{4.67}$$

A bar placed above the symbol for some property of the conduction electrons, such as velocity, means that its value is averaged over a cyclotron orbit, e. g.,

$$\bar{\mathbf{v}} = \frac{1}{2\pi} \int_0^{2\pi} \mathbf{v} \, d\Theta. \tag{4.68}$$

The electron force $K_i = \partial \tau_{ij}^d / \partial x_j$ acting upon the lattice can be obtained by varying the electron energy [equation (4.64)] with respect to the displacement $\mathbf{u}$ and averaging over the conduction electrons.

The electric current density $\mathbf{j}$ and the tensions due to the nonequilibrium electrons (or dynamic tensions of electronic origin), $\tau_{ij}^d$,

$$\mathbf{j} = \frac{2e}{h^3} \int_{FS} \mathbf{v} f \, d^3 p, \qquad \tau_{ij}^d = \frac{2}{h^3} \int_{FS} \Lambda_{ij} f \, d^3 p, \tag{4.69}$$

now may be expressed as surface integrals over the FS,

$$j_i = \frac{2e}{h^3} \dot{\varepsilon}_{kl} \int_{FS} |m_c|\, dp_z \int_{FS} v_i \tilde{\Lambda}_{kl}\, d\Theta$$

$$+ \frac{2e^2}{h^3} F_k \int_{FS} |m_c|\, dp_z \int_{FS} v_i \tilde{v}_k\, d\Theta$$

$$\equiv \beta_{i,kl} \dot{\varepsilon}_{kl} + \sigma_{ik} F_k, \tag{4.70}$$

$$\tau_{ij}^d = \frac{2}{h^3} \dot{\varepsilon}_{kl} \int_{FS} |m_c|\, dp_z \int_{FS} \Lambda_{ij} \tilde{\Lambda}_{kl}\, d\Theta$$

$$+ \frac{2e}{h^3} F_k \int_{FS} |m_c|\, dp_z \int_{FS} \Lambda_{ij} \tilde{v}_k\, d\Theta$$

$$\equiv \alpha_{ijkl} \dot{\varepsilon}_{kl} + \beta_{ij,k} F_k, \tag{4.71}$$

where the variables with a tilde are defined by equation (4.66). The electroacoustic coefficients, consisting of the $\alpha_{ijkl}$ that define the conduction electron contributions to the elasticity of a metal, the electrical conductivity $\sigma_{ij}$, and the deformation conductivities $\beta_{i,kl}$ and $\beta_{ij,k}$, can be written then in a more convenient form for future use. We define

$$< ab > \equiv \int_{FS} |m_c|\, dp_z \int_{FS} \widetilde{ab}\, d\Theta$$

$$= \int_{FS} |m_c|\, dp_z \sum_{n=-\infty}^{\infty} \frac{a_n b_n^*}{\nu - i\,(\mathbf{k}\cdot\bar{\mathbf{v}} - \omega - n\Omega)}, \tag{4.72}$$

where * denotes the complex conjugate and

$$a_n = \frac{1}{2\pi} \int_0^{2\pi} d\Theta\, a(\Theta) \exp\left[ -\frac{i}{\Omega} \int_0^\Theta \{\mathbf{k}\cdot [\mathbf{v}\,(\Theta_1) - \bar{\mathbf{v}}] + n\Omega\}\, d\Theta_1 \right]. \tag{4.73}$$

As a result,

$$\beta_{i,kl} = \frac{2e}{h^3} < v_i \Lambda_{kl} >, \qquad \sigma_{ik} = \frac{2e^2}{h^3} < v_i v_k >,$$

$$\alpha_{ijkl} = \frac{2}{h^3} < \Lambda_{ij} \Lambda_{kl} >, \qquad \beta_{ik,l} = \frac{2e}{h^3} < \Lambda_{ik} v_l > . \tag{4.74}$$

The material equations (4.70) and (4.71) are the generalized Ohm's and Hooke's laws, and may be regarded as supplements to Maxwell's equation

$$\nabla \times (\nabla \times \mathbf{E}) = -\frac{4\pi}{c} \frac{\partial \mathbf{j}}{\partial t}, \tag{4.75}$$

where the term proportional to $\partial \mathbf{E}/\partial t$ has been neglected, and the equations of elasticity theory,

$$\rho \ddot{u}_i = c_{ijkl} \frac{\partial \varepsilon_{kl}}{\partial x_j} + \frac{\partial \tau_{ij}^d}{\partial x_j}, \tag{4.76}$$

where the $c_{ijkl}$ are quasi-static elastic moduli that contain the contributions of the lattice and the equilibrium electrons.

Next we will make some simplifications since our interest lies in investigating the acoustic analog of the Faraday effect, the only one of the MPP observed in nonmagnetic metals. Thus we will restrict ourselves to $\mathbf{H} \parallel \mathbf{k} \parallel C_4$, where $C_4$ is a fourth-order rotational symmetry crystal axis. Since the eigenmodes for this geometry are circularly polarized, we will introduce circular components of vectors (e.g., $u^{\pm} = u_1 \pm iu_2$, etc.) as was done in Sec. 4.3.2. This definition does not refer to $k^{\pm}$, which is used for the $z$ components of the wave vectors of circularly polarized modes: $\mathbf{k} = (0, 0, k^{\pm})$.

Recall that the Faraday effect requires a crystal axis symmetry that is at least three-fold; a development similar to that given below can be carried out for a trigonal or a hexagonal crystal axis.

If the solutions of the system of equations (4.75)–(4.76) are expressed in the form $\mathbf{u}, \mathbf{E} \sim \exp\left[i\left(\omega t - \mathbf{k} \cdot \mathbf{r}\right)\right]$, the determinant of the system, when set equal to zero, will give the dispersion equations for the electromagnetic and elastic eigenmodes in a metal. Another way of solving the equations is to Fourier-transform them to obtain a set of homogeneous linear equations and, again, set the determinant equal to zero to get the same result, namely,

$$
\left(\frac{k^{\pm 2}}{k_s^{\pm 2}} - 1\right)\left(\frac{k_E^{\pm 2}}{k^{\pm 2}} - 1\right) = \frac{4\pi i}{\rho \omega c^2}\left(\beta_1^{\pm} \pm \frac{icHk^{\pm}}{4\pi\omega}\right)\left(\beta_2^{\pm} \pm \frac{icHk^{\pm}}{4\pi\omega}\right),
$$

$$(4.77)$$

where

$$
k_s^{\pm 2} = \left(\rho\omega^2\right)\left(c_{1313} + i\omega\alpha^{\pm} + \frac{H^2}{4\pi}\right)^{-1}, \tag{4.78}
$$

$$
k_E^{\pm 2} = -i\left(4\pi\omega\sigma^{\pm}/c^2\right), \tag{4.79}
$$

and the circular components of the electroacoustic coefficients are defined as

$$
\alpha^{\pm} = \alpha_{1313} \pm i\alpha_{1323}, \qquad \sigma^{\pm} = \sigma_{11} \pm i\sigma_{21}. \tag{4.80}
$$

In addition, if $\mathbf{k}$ lies in a plane of mirror symmetry as in the present case, $\beta_1^{\pm} = \beta_2^{\pm} \equiv \beta^{\pm}$ (as proved in [98]), so

$$
\beta^{\pm} = \beta_{113} \pm i\beta_{123}. \tag{4.81}
$$

The right-hand side of equation (4.77) can be regarded as the dimensionless parameter of coupling between the elastic and electromagnetic systems, $K_c$. If $K_c$ vanishes, equation (4.77) transforms into two equations, $k^{\pm 2} - k_s^{\pm 2} = 0$ and $k^{\pm 2} - k_E^{\pm 2} = 0$, which describe the dispersion of pure elastic and pure electromagnetic waves, respectively. If the coupling parameter is finite, there will be solutions corresponding to the dispersion of coupled acousto-electromagnetic modes, i.e., waves characterized by both elastic displacements and electromagnetic fields. We will begin a complete treatment of these waves by first discussing pure electromagnetic waves in metals.

### 4.4.2  Electromagnetic Waves

The dispersion of electromagnetic waves is completely determined by the circular components of the conductivity tensor, specifically its dissipative $\sigma_{11}$ and nondissipative (Hall) $\sigma_{21}$ parts. In general, the conductivity depends on the magnetic field and the wave vector. However, in the so-called local regime, which corresponds formally to the condition

$$\left| \frac{\mathbf{k} \cdot \bar{\mathbf{v}}}{\Omega} \right| \ll 1, \tag{4.82}$$

the dependence on $\mathbf{k}$ may be neglected and the dispersion of the electromagnetic waves is given by

$$\left( k^{\pm} \right)^2 = \mp \frac{4\pi |e| (N_e - N_h) \omega}{cH} \left( 1 \mp i\gamma \right), \tag{4.83}$$

where $\gamma = 1/\Omega\tau$, and $N_e$ and $N_h$ are the electron and hole concentrations in the metal.

It can be seen that, if $N_e \neq N_h$ and

$$|\Omega\tau| \gg 1 \tag{4.84}$$

for one of the polarizations [e.g., $(-)$ for an electron-type conductor], the wave vector will be essentially real. In other words, in a strong magnetic field a noncompensated metal can propagate circularly polarized electromagnetic waves. The sign of the circular polarization depends on the dominant charge carriers and the direction of the applied magnetic field. These waves, called helicons, were predicted by Agrain [99] and by Konstantinov and Perel' [100]. They were discovered in sodium by Bowers et al. [101] and later observed in many metals. The results of both theoretical and experimental investigations are published in review papers [102–105].

For the inequality opposite to that given in (4.82), namely

$$|\mathbf{k} \cdot \bar{\mathbf{v}}_{max}| \geq \Omega, \tag{4.85}$$

where $\bar{\mathbf{v}}_{max}$ is the maximum value of the average electron velocity, Doppler-shifted cyclotron resonance (DSCR)occurs, which is, in fact, one of the manifestations of spatial dispersion.

Note that all of the expressions for the electroacoustic coefficients (4.72)–(4.74) have a resonant denominator, $\nu - i\,(\mathbf{k}\cdot\bar{\mathbf{v}} - \omega - n\Omega)$. In the collisionless limit (i.e., for $\nu \to 0$), the vanishing of this denominator corresponds to the condition for cyclotron resonance of the electrons traveling along $\mathbf{H}$ with average velocity $\bar{\mathbf{v}}$. Since the electrons in a metal have a quasi-continuous but finite range of $m_c$ and $\bar{\mathbf{v}}$ values, there is a region of cyclotron absorption extending from $H_{min}$ to $H_K$. The lower limit is due to the requirement that $\Omega\tau > 1$, whereas the upper corresponds to the Kjeldaas field (or Kjeldaas edge for DSCR) [106] defined as

$$H_K = \left| \frac{\mathbf{k}\cdot\bar{\mathbf{v}}_{max}\,(m_c)_{max}\,c}{e} \right|, \tag{4.86}$$

above which the condition for cyclotron resonance cannot be fulfilled.

In ultrasonic experiments DSCR manifests itself as a nonmonotonic, sometimes resonant dependence of absorption and phase velocity on $\omega$ or $H$ (see the reviews [70] and [71]).

Taking into account the spatial dispersion of $\sigma^{\pm}$ under DSCR conditions leads to the following: The quadratic behavior ($\omega \sim k^2$) typical of helicon dispersion in the long-wavelength limit becomes distorted as it approaches the region near $\mathbf{k} \cdot \mathbf{v}_{max} = \Omega$, and then is transformed into a new branch of the dispersion curve.

This idea was formulated by McGroddy et al. [107] and Overhauser and Rodriguez [108]. They considered a metal with a spherical FS and noted a narrow region close to but above the Kjeldaas field in which an electromagnetic wave can exist and called it a Doppler-shifted cyclotron mode.

Antoniewicz [109, 110] studied a more complicated FS model, namely, one which has a concave portion in its cross-sectional area. He pointed out that, for magnetic fields below the conventional Kjeldaas edge, there exists a "window" where an electromagnetic wave of a helical type propagates with small damping.

Kaner and Skobov [111] predicted that a wave of this type could occur in compensated metals (i.e., in metals with $N_h = N_e$) below the Kjeldaas edge, while Konstantinov and Skobov [112] completed a thorough theoretical study of these waves and called them dopplerons.

Contrary to the situation for helicons, doppleron dispersion depends essentially on local features of the FS of a metal. In particular, it requires charge carriers with extremal magnitudes of $m_c\bar{v}_z = (1/2\pi)\,\partial S/\partial p_z$ and depends on the number of such carriers. Therefore, dopplerons exist in compensated metals as well as in non-compensated ones.

The first experimental observation of dopplerons was made in copper and reported by Antoniewicz et al. [113]. More detailed results of the investigation were published by Wood and Gavenda [114]. Later, waves of this type were found in cadmium [115–118], indium [119], aluminum [120], and tungsten and molybdenum [121], and investigated further in copper [122].

As examples of helicon and doppleron dispersion, Fig. 4.3 shows typical

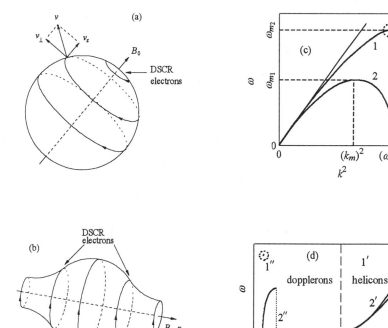

FIGURE 4.3. (a) Spherical and (b) corrugated-cylindrical Fermi surfaces. (c) Dispersion Curve 1 is for the spherical FS, while 2 is for the corrugated-cylindrical FS (the straight line is for the local theory.) (d) Dispersion curves for helicons (1′ and 2′) and dopplerons (1″ and 2″). Curves 1′, and 1″ relate to the FS in (a), while Curves 2′ and 2″ relate to the FS in (b). The small segment 1″ comes from the upper portion of Curve 1 indicated by the dotted circle in (c). (After Fig. 4 in [105])

curves given in the review by Petrashov [105]. The curves were obtained for two model FSs: a sphere and a corrugated cylinder. Taking account of nonlocal effects due to DSCR leads to the conclusion that the main contribution is from the electrons located near the limit point of the sphere, or on the belt of electrons around the FS which corresponds to the extremum of the derivative $\partial S/\partial p_z$ for the cylinder. This belt of effective electrons is situated between the cross sections of the FS with minimum and maximum magnitudes of $S$. Since the total number of electrons near the limit point of the sphere and the sizes of the transverse components of the velocity are small, the contribution of these carriers to $\sigma^\pm$ is also small. This circumstance results in a very narrow region where dopplerons can exist. On the other hand, the contribution of the effective electrons on the corrugated cylinder is considerable, so the region of doppleron existence is considerably larger.

An important peculiarity of the dopplerons for these models is the fact that the phase $[\mathbf{v}_E = (\omega/k^2)\mathbf{k}]$ and group $(\mathbf{g}_E = \partial\omega/\partial\mathbf{k})$ velocities of these electromagnetic waves have opposite directions. This feature is represented in Fig. 4.3 by different signs of $k$ relating to the dispersion of helicons and dopplerons, while $\mathbf{g}_E$ is assumed to be positive in all cases. However, there can be other FS shapes which result in dopplerons with a common direction of both velocities. Dopplerons of this type have been observed in aluminum [120] and indium [37]. A more complete discussion of dopplerons can be found in [123].

## Summary

Finally, a few words to clarify the conditions necessary for low-frequency (compared to the plasma frequency) waves to propagate in a metal. As shown in this section, the wave vectors of circularly polarized electromagnetic modes depend on $\sigma_{11}$ and $\sigma_{21}$, which determine the ordinary current $\mathbf{j}_J$ and the Hall current $\mathbf{j}_H$, respectively. Note that $\mathbf{j}_J$ produces Joule heat, whereas $\mathbf{j}_H$ has a nondissipative nature because $\mathbf{j}_H \perp \mathbf{E}$. Helicons and dopplerons can be observed only if $\sigma_{21} > \sigma_{11}$ and, from this point of view, they are waves of the Hall current. Their existence is as natural as the propagation of nondissipative waves of displacement current in dielectrics.

## 4.4.3  Coupled Electromagnetic and Elastic Waves

### Helicon–phonon modes

In discussing the interaction of elastic and electromagnetic subsystems, we will begin with helicon–phonon resonance (HPR). The theory of this phenomenon was developed by a number of authors [33, 95, 124–127]. The first experiments on coupled helicon–phonon modes were done in potassium using electromagnetic excitation of the waves [27], while in [128] a purely ultrasonic technique was used. Later on, the data obtained in these experiments were compared with theoretical calculations [129–131].

In order to reveal the principal features, we will consider the local regime (i.e., when there is no spatial dispersion) for $\sigma^{\pm}$ and assume that the other electroacoustic coefficients vanish. Thus, by defining $k_E^{\pm 2}$ with equation (4.79) and using the static value of $\sigma^{\pm}$, we can transform equation (4.77) into

$$\left(k^{\pm 2} - k_0^2\right)\left(k^{\pm 2} - k_E^{\pm 2}\right) - \eta k^{\pm 2} k_E^{\pm 2} = 0, \tag{4.87}$$

where $k_0 = \omega/s_0$, $s_0$ is the phase velocity of transverse elastic waves, defined with the quasi-static elasticity modulus $c_{1313}$ at $H = 0$, and

$$\eta = \frac{H^2}{4\pi c_{1313}} \tag{4.88}$$

is the coupling parameter, defined in the local regime. The roots of equation (4.87) are

$$k_{1,2}^{\pm\;2} = \frac{k_0^2 + k_E^{\pm 2}\left(1 + \eta\right) \pm \left[\left(k_0^2 + k_E^{\pm 2}\left(1 + \eta\right)\right)^2 - 4k_0^2 k_E^{\pm 2}\right]^{\frac{1}{2}}}{2}. \tag{4.89}$$

Three solutions of this equation yield positive values of $k^2$. Two of them are obtained for the polarization corresponding to a helicon and represent coupled helicon–phonon modes (see Fig. 4.4), while the solution with the opposite polarization corresponds to a pure elastic wave.

The dispersion curves presented in Fig. 4.4 display a gap resulting from mode coupling between the two branches. Such a situation will occur only if the dimensionless coupling parameter exceeds the losses at the resonance $(H \simeq H_r)$:

$$H_r^2/4\pi\rho s_0^2 \gg |\gamma|. \tag{4.90}$$

The resonant field is determined by the crossing of the dispersion curves of the non-interacting helicon and elastic modes, and is given by

$$H_r = \frac{4\pi s_0^2 \left|(N_e - N_h)\,e\right|}{\omega c}. \tag{4.91}$$

If the coupling parameter is small there is no gap between the dispersion curves. This situation is illustrated in Fig. 4.5. In this case, the coupled waves are called helicon-like and elastic- or phonon-like modes, since one type of energy (electromagnetic or elastic) transmitted by the wave is dominant.

In the case of DSCR it should be noted that the dissipative component of the nonlocal conductivity and also the absorption of the elastic waves are

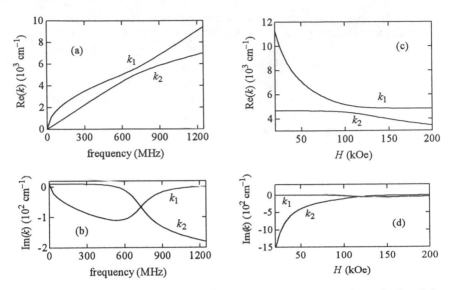

FIGURE 4.4. Wave numbers of coupled helicon–phonon modes calculated for indium with $\mathbf{H} \parallel [001]$, using $s_0 = 0.95 \times 10^5$ cm/sec, $N_h - N_e = 3.93 \times 10^{22}$ cm$^{-3}$, $\tau = 10^{-11}$ s, and $m = 9.11 \times 10^{-31}$ kg. (a) and (b) Real and imaginary parts, respectively, for $H = 100$ kOe; (c) and (d) Real and imaginary parts, respectively, for $\omega/2\pi = 100$ MHz.

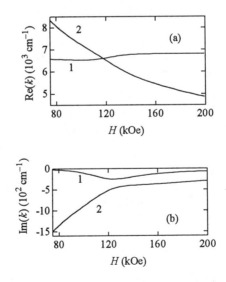

FIGURE 4.5. (a) The real and (b) the imaginary parts of the wave numbers of coupled elastic-like (1) and helicon-like (2) modes as functions of the magnetic field for $\omega/2\pi = 100$ MHz, calculated with the parameters used in the previous figure, except $\tau = 2 \times 10^{-12}$ sec.

significant even in the collisionless limit, thus the weak-coupling condition is more common in this field range.

With the assumption that the contributions of the terms originating from the electroacoustic coefficients are small with respect to $k_0$, equation (4.77) may be solved by iteration. The zero-order approximation for the branch representing the elastic-like mode (i.e., the mode for which the elastic energy prevails) is given by $k_e^\pm = k_0$. In fact, this is the wave number of a pure elastic mode in a metal when the influence of the magnetic field and non-equilibrium electrons may be neglected.

The first iteration gives an equation in which all of the electroacoustic coefficients, $\alpha^\pm$, $\beta^\pm$, and $\sigma^\pm$ [the latter is included in $k_E^\pm$ according to (4.79)], are functions of $k_0$:

$$k_e^\pm = k_0 \left\{ 1 - \frac{i\omega\alpha^\pm + \left(H^2/4\pi\right)}{2c_{1313}} - \left(\frac{2\pi k_0^2}{\rho c^2}\right) \frac{\left[\beta^\pm \pm i\left(cHk_0/4\pi\omega\right)\right]^2}{k_0^2 - k_E^2} \right\}.$$

$$(4.92)$$

Here the term proportional to $\alpha^\pm$ describes the dynamic effects caused by the elasticity of the electron gas in a metal and is usually called the deformation term. The last term in equation (4.92) describes the influence of the electromagnetic fields accompanying elastic wave propagation. It is called the electromagnetic (or field) term and has the form of a coupling parameter divided by a resonant denominator. This denominator intensifies the contribution of the field term when the real part of the wave number for the weakly damped electromagnetic mode (in the DSCR region, a doppleron) becomes comparable with the wave number of the elastic wave. Note that in a metal with an anisotropic FS the coupling parameter may also have a resonant character when DSCR occurs. Thus, in many metals, the contribution of the electromagnetic field term to ultrasonic absorption is dominant at low temperatures.

### Doppleron–phonon modes

In discussing the effects caused by spatial dispersion, it is important to note that the shape of the FS strongly influences the electromagnetic wave spectrum (dopplerons). Owing to the complicated dependence of conductivity on $k$, there may be some roots of the dispersion equation related to weakly damped, circularly polarized modes. Some of them correspond to waves having phase and group velocities with parallel directions, but more typically the energy propagation and the wave vector are in opposite directions. In this case, if the coupling parameter is large and the losses are small, there will be a gap in the spectrum of coupled doppleron–phonon modes: some interval of $\omega$ or $H$ becomes forbidden for the wave excitation. This feature leads to an appearance of specific resonant characteristics in the dependences of $\phi$ and $\varepsilon$ on the frequency or magnetic field.

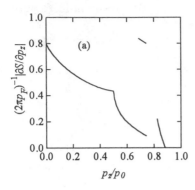 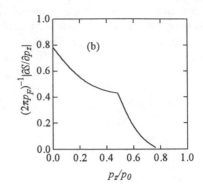

FIGURE 4.6. Derivatives of the cross-sectional area of the FS of indium in the second Brillouin zone with $p_z \parallel [001]$: (a) almost-free electron model; (b) model discussed here. (After [119].)

To illustrate the most typical features of doppleron–phonon resonance (DPR), we will use the approach and model that were introduced in [119] for describing dopplerons in indium. When simulating doppleron dispersion, particular attention should be given to modeling not the FS itself, but the derivative $\partial S/\partial p_z \sim m_c \bar{v}_z$. It is this product that determines the position of DSCR features when electroacoustic coefficients (in our case, components of the conductivity tensor) are plotted against $\omega$ or $H$.

In the paper mentioned above, $\partial S/\partial p_z$ was approximated with two parabolas for the FS in the second Brillouin zone (see Fig. 4.6), whereas a small group of negative charge carriers in the third zone was neglected. Thus, $\partial S/\partial p_z$ is obtained from

$$S(p_z) = 2\pi p_0 p_F \left[ A + B + y_3 \left( x_1 - \frac{|p_z|}{p_0} \right) + \frac{y_4 - y_3}{3x_1^2} \left( x_1 - \frac{|p_z|}{p_0} \right)^3 \right]$$

$$(4.93)$$

for $|p_z| \leq p_0 x_1$, and

$$S(p_z) = 2\pi p_0 p_F \left[ B + \frac{y_3}{3 \left( x_2 - x_1 \right)^2} \left( x_2 - \frac{|p_z|}{p_0} \right)^3 \right] \qquad (4.94)$$

for $p_0 x_1 \leq |p_z| \leq p_0 x_2$, where $2p_0$ is the dimension of the Brillouin zone along the direction [100], $p_F = 1.1 p_0 = 1.6 \times 10^{-19}$ g cm/sec, $x_1 = 0.48$, $x_2 = 0.8$, $y_3 = 0.43$, $y_4 = 0.8$, $A = y_3(x_2 - x_1)/3$, and $B$ is determined by the requirement of equality of the nonlocal conductivity $\sigma_{21}$ to the Hall conductivity as $q \to 0$ ($q$ is the parameter of nonlocality and $q = 0$ corresponds to the local limit). This requirement may be written as

$$N_h = \frac{4}{h^3} \int_0^{p_0 x_2} S(p_z) \, dp_z. \qquad (4.95)$$

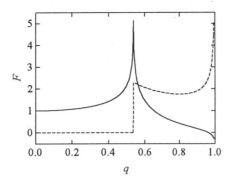

FIGURE 4.7. Normalized nonlocal conductivity as a function of $q$ in the collisionless limit. Solid line: $\mathrm{Re}(F)$. Dashed line: $\mathrm{Im}(F)$.

The size of the hole concentration $N_h$ was taken to be $3.93 \times 10^{22}\,\mathrm{cm}^{-3}$ [132]. The parameter of nonlocality is defined as the ratio of the displacement of the effective carriers along $\mathbf{H}$ during a cyclotron period $T_c = 2\pi/\Omega$ to the wavelength. Here the effective holes have $p_z = p_0 x_1$, thus

$$q = \frac{k c p_F y_3}{|e| H}. \tag{4.96}$$

In the collisionless limit, the circular components of the conductivity are

$$\sigma^{\pm} = i \frac{|e| c N_h}{H} \left[ \pm \mathrm{Re} F(q) + i \mathrm{Im} F(q) \right], \tag{4.97}$$

with the normalized nonlocal conductivity

$$
\begin{aligned}
F(q) = \frac{n_0}{N_h} \Bigg\{ & \frac{x_1^2 y_3}{3(\xi - 1) q} \left[ \ln \frac{(1 - q)(1 + \xi q)}{(1 + q)(1 - \xi q)} + \frac{1}{2q} \ln \frac{1 - q^2}{1 - \xi^2 q^2} \right] \\
& + \frac{x_1(A + B)}{2\sqrt{\xi - 1}} \left[ \frac{2}{\sqrt{q(1 + q)}} \tan^{-1} \sqrt{\frac{(\xi - 1) q}{1 + q}} \right. \\
& \left. \qquad\qquad + \frac{1}{\sqrt{q(1 - q)}} \ln \frac{\sqrt{1 - q} + \sqrt{(\xi - 1) q}}{\sqrt{1 - q} - \sqrt{(\xi - 1) q}} \right] \\
& + B \frac{x_2 - x_1}{\sqrt{q}} \left( \tan^{-1} \sqrt{q} + \frac{1}{2} \ln \frac{1 + \sqrt{q}}{1 - \sqrt{q}} \right) \\
& - \frac{(x_2 - x_1)^2 y_3}{6 q^2} \ln(1 - q^2) \Bigg\},
\end{aligned}
\tag{4.98}
$$

where $n_0 = (4\pi p_0^2 p_F)/h^3$ and $\xi = y_4/y_3$. As shown in Fig. 4.7, $F(q)$ is real for $0 \leq q \leq 1/\xi$ and complex for $q > 1/\xi$, where $1/\xi = 0.54$ for the model under consideration.

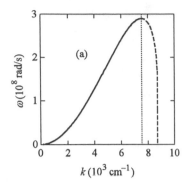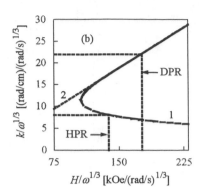

FIGURE 4.8. (a) Dispersion of helicons (left of vertical line) and dopplerons (right of vertical line) for (+) polarization in the model under discussion at $H = 70$ kOe. (b) Magnetic field dependence of the wave number of the helicon–doppleron branch (Curve 1). Line 2 shows the asymptote for the short wavelength doppleron region, $q \to 1/\xi$.

The dispersion equation $k^{\pm^2} = k_E^{\pm^2}$ can be written as

$$\pm \frac{\omega_0}{\omega} = \frac{F(q)}{q^2}, \tag{4.99}$$

where $(\pm)$ refer to the $(\pm)$ polarizations and

$$\omega_0 = \frac{|e|H^3}{4\pi c p_F^2 y_3^2 N_h}. \tag{4.100}$$

For the model under discussion, equation (4.99) gives three roots, which correspond to weakly damped modes: helicons and dopplerons for (+) polarization $(q < 1/\xi)$, and dopplerons for $(-)$ polarization $(q > 1)$. For the present we will consider only (+)-polarized waves. The dispersion curve and the magnetic field dependence of the wave number for helicons and dopplerons $(k/\omega^{\frac{1}{3}}$ as a function of $H/\omega^{\frac{1}{3}})$ are presented in Fig. 4.8. As can be seen, the helicon transforms into a doppleron when the local regime is replaced by the nonlocal one. Thus, this type of electromagnetic wave may be regarded as arising from an unseparated helicon–doppleron branch. Actually, there is a particular point that separates the two components: the highest point on the dispersion curve. The left part corresponds to the helicon-like solution in spite of some distortions of the quadratic dependence caused by DSCR effects, whereas the right part relates to the doppleron-like solution, entirely initiated by DSCR. The group velocity $g_E = \partial\omega/\partial k$ for the helicon-like solution is positive while that for the doppleron-like is negative, with $k > 0$ in both cases.

If propagation of the waves in the positive $z$ direction is under consideration, the helicon should be regarded as a wave with positive $k$, contrary to the case for the doppleron. The latter must have negative $k$ to provide a positive $g_E$.

The value of $H$ for DPR (as well as for HPR) may be determined by reference to Fig. 4.8(b). The phonon resonance condition corresponds to the projection to the $H/\omega^{\frac{1}{3}}$ axis of the intersection of Curve 1 and the horizontal dashed line $k/\omega^{\frac{1}{3}} = $ const, where $k = k_0$. It is clear that either DPR or HPR can occur at a given frequency, as suggested by the two examples in the figure. If there are more doppleron-like solutions of the dispersion equation, at a given $\omega$ one can observe HPR as well as DPR associated with these dopplerons.

The dispersion curves for the coupled waves may be obtained with the use of equation (4.77) accompanied by some simplifications. The simplifications are based on the vanishing of the electroacoustic coefficients $\alpha^{\pm}$ and $\beta^{\pm}$ above the Kjeldaas edge in our model (i.e., in the region where $q < 1/\xi$, just where the helicon–doppleron branch is defined). Thus $(k_s^{\pm})^2$ can be replaced by $\rho\omega^2/c_{1313} \equiv k_0$ and the dispersion equation expressed in the following form for $(+)$ polarization:

$$i\frac{4\pi\sigma}{c^2 s^2}\omega^3 + \frac{k^2}{s^2}\omega^2 - i\left(\eta + 1\right)\frac{4\pi\sigma}{c^2}k^2\omega - k^4 = 0. \qquad (4.101)$$

Recall that $\sigma$ is a function of $k$ and $H$, defined by equation (4.97), but not of $\omega$, since the frequency dependence of the electroacoustic coefficients may be neglected for $\omega\tau < 1$.

This equation has three solutions. Two of them, shown in Fig. 4.9(a), relate to typical coupled modes. It may easily be seen that under strong coupling conditions (e.g., in the collisionless approximation) both the helicon–phonon and the doppleron–phonon resonances cause a gap to appear in the dispersion curves near the point where they cross. This gap appears near the point where the formerly pure electromagnetic and elastic wave dispersion curves cross. Nevertheless, these two gaps are quite different.

At the frequency corresponding to HPR for a particular magnetic field one may find three $(+)$-polarized waves: a pure electromagnetic mode (doppleron) and two coupled modes (helicon–phonon or helicon-like, and phonon–helicon or phonon-like.)

Under DPR the gap is formed in such a way that in some interval of $\omega$ only the helicon-like mode can exist. This important distinction arises from the fact that the group velocities of the doppleron discussed here and the elastic wave have opposite directions (for a particular $k$). It may be shown that, if the doppleron also has a positive group velocity, the gap which appears is of the HPR type. The curves presented in Fig. 4.9(b) also illustrate a distinguishing feature of DPR: there is no solution corresponding to a doppleron–phonon mode in the resonant interval of $H$, contrary to the situation that holds for HPR [see Fig. 4.4(c)].

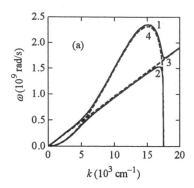 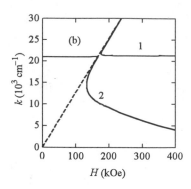

FIGURE 4.9. (a) Dispersion of coupled helicon–phonon and doppleron–phonon modes (Curves 1 and 2) for $H = 140\,\mathrm{kOe}$ with strong coupling. Dashed line 3 relates to a pure elastic wave, while dashed curve 4 relates to a pure helicon–doppleron wave. (b) Magnetic field dependences of the wave numbers corresponding to the low-damping doppleron–phonon modes (Curves 1 and 2) for $\omega = 2 \times 10^9\,\mathrm{rad/s}$. The dashed line represents the cyclotron absorption edge.

## 4.5   Bulk Phenomena: The Faraday and Cotton–Mouton Effects

In subsections (2.2) and (2.3) we presented methods of measuring the ellipticity and rotation of the polarization in terms of the amplitude and phase of the signal propagated through the specimen for the acoustic analogs of the Faraday and Cotton–Mouton effects. Here we will discuss equations involving such physical parameters as wave numbers, phase velocities, absorption, transmission coefficients, etc., which determine the values of $\varepsilon$ and $\phi$ in an experimental investigation of a bulk polarization phenomenon.

### 4.5.1   Weak Coupling

The Faraday effect

A typical arrangement for an experimental investigation of polarization phenomena is given in Fig. 1.1. Piezoelectric transducers are located on two parallel faces of a specimen; one of the transducers is connected to a high-frequency generator that produces a constant amplitude voltage, while the second is connected to a receiver (e.g., as shown in Fig. 3.3). The first transducer generates linearly polarized elastic displacements of the surface to which it is attached that then propagate through the specimen to the second transducer, which is sensitive to displacements along a particular direction. It transforms the elastic displacements into a voltage that is sent to the input of a receiver.

Under weak-coupling conditions, the first transducer generates two circularly polarized elastic-like modes (the proof of this statement will be given later) with equal amplitudes. Thus, equations (2.5) may be written as

$$\varepsilon = \frac{e^{\operatorname{Im}\Delta k_e^+ z} - e^{\operatorname{Im}\Delta k_e^- z}}{e^{\operatorname{Im}\Delta k_e^+ z} + e^{\operatorname{Im}\Delta k_e^- z}} = \frac{e^{-\Delta\Gamma_e^+ z} - e^{-\Delta\Gamma_e^- z}}{e^{-\Delta\Gamma_e^+ z} + e^{-\Delta\Gamma_e^- z}}, \tag{4.102}$$

$$\phi = \frac{1}{2}\operatorname{Re}\left(\Delta k_e^+ - \Delta k_e^-\right) z = \frac{k_0}{2}\left(\frac{\Delta s_e^-}{s_0} - \frac{\Delta s_e^+}{s_0}\right) z, \tag{4.103}$$

where $\Delta\Gamma_e^\pm = \Gamma_e^\pm(H) - \Gamma_0$, $\Delta k_e^\pm = k_e^\pm(H) - k_0$, and $\Delta s_e = s_e^\pm(H) - s_0$, and where $\Gamma_0$, $k_0$, and $s_0$ are the values for $H = 0$ independent of the polarization. Note that the imaginary component of a wave number, $\operatorname{Im} k_e = -\Gamma_e$, is always taken to be negative to ensure that the wave amplitude decays when propagating in the positive direction of the $z$ axis.

### The Cotton–Mouton effect

For this case the eigenmodes are waves of mutually perpendicular linear polarization and the first transducer is usually mounted in such a way that two normal modes are generated with equal amplitudes and initial phases. Thus we can use equation (2.4) with $\varphi_{1,2}^\pm = \pm\pi/4$ to obtain the following result for the parameter $p$ defined in equation (2.6):

$$p = \frac{e^{-\Delta\Gamma_1 z + i(\operatorname{Re}\Delta k_1 z - \pi/4)} + e^{-\Delta\Gamma_2 z + i(\operatorname{Re}\Delta k_2 z + \pi/4)}}{e^{-\Delta\Gamma_1 z + i(\operatorname{Re}\Delta k_1 z + \pi/4)} + e^{-\Delta\Gamma_2 z + i(\operatorname{Re}\Delta k_2 z - \pi/4)}}, \tag{4.104}$$

or

$$|p|^2 = \frac{4\cos^2\left[(\Delta k_2 - \Delta k_1)\, z\right] e^{-2(\Delta\Gamma_2 + \Delta\Gamma_1)z} - \left(e^{-2\Delta\Gamma_2 z} - e^{-2\Delta\Gamma_1 z}\right)^2}{\left\{e^{-2\Delta\Gamma_2 z} - 2\sin\left[(\Delta k_2 - \Delta k_1)\, z\right] e^{-(\Delta\Gamma_2 + \Delta\Gamma_1)z} + e^{-2\Delta\Gamma_1 z}\right\}^2}, \tag{4.105}$$

which we can then insert into the definitions of $\varepsilon$ and $\phi$ in equation (2.7) to obtain

$$\phi = -\frac{1}{2}\tan^{-1}\frac{e^{-2\Delta\Gamma_2 z} - e^{-2\Delta\Gamma_1 z}}{2\cos\left[(\Delta k_2 - \Delta k_1)\, z\right] e^{-(\Delta\Gamma_2 + \Delta\Gamma_1)z}}, \tag{4.106}$$

where $\Delta\Gamma_{1,2} = \Gamma_{1,2}(H) - \Gamma_0$, $\Delta k_{1,2} = k_{1,2}(H) - k_0$ are the magnetic field variations of the absorption and wave numbers of the normal modes designated by the indexes 1 and 2.

It can easily be seen that in this case both $\varepsilon$ and $\phi$ depend on the magnetic field variations of both the absorptions and the phase velocities of the normal modes, contrary to the situation for the weak-coupling Faraday effect, where $\varepsilon$ is associated with dichroism and $\phi$ with birefringence [see equations (4.102) and (4.103)].

## 4.5.2  Strong Coupling

When there is strong mode coupling, the preceding discussion becomes more complicated because three shear modes are generated by the transducer. Two of them have one of the polarizations and one has the other. Thus, the ellipticity and rotation of the polarization are the result of the interference of three waves that propagate the elastic vibrations, namely, two strongly coupled modes and an elastic-like mode of nonresonant polarization. The latter may be regarded as a pure elastic wave, since the energy of the electromagnetic field transported by the wave, although not zero, is, in fact, very small in comparison with the elastic energy.

By analogy with birefringence and dichroism in optics, such terms as trirefringence and trichroism may be introduced in the discussion of phenomena of this type. Up to now only the acoustic analog of the Faraday effect has been investigated for strong coupling, so only this phenomenon will be discussed.

The spatial dependences of the elastic displacements may be written either as

$$u^r(z) = u_1 e^{-ik_1^r z} + u_2 e^{-ik_2^r z}, \qquad (4.107)$$

$$u^n(z) = u_0 e^{-ik_1^n z}, \qquad (4.108)$$

where superscripts $r$ and $n$ designate resonant and nonresonant polarizations. The parameter $p$ should be written as

$$p = \frac{u^r(z)}{u^n(z)} \quad \text{or} \quad p = \frac{u^n(z)}{u^r(z)}, \qquad (4.109)$$

depending on which polarization is resonant: The first form applies when the $(-)$ polarization is resonant, and the second when the $(+)$ one is.

The coefficients $u_1$ and $u_2$ satisfy the requirement that $u_1 + u_2 = u_0$ (based on the linear polarization of the elastic waves sent through the specimen by a generating transducer located at $z = 0$). In addition, one can speculate (although this has not been proved) that

$$\frac{u_1}{u_2} = \frac{D_{uu}^{(1)}}{D_{uu}^{(2)}}, \qquad (4.110)$$

where the $D_{uu}^{(i)}$ are the amplitude transmission coefficients through the interface between dielectric and metallic media for an elastic wave of the resonant polarization associated with the $i^{th}$ mode (i.e., the $i^{th}$ solution of the dispersion equation for the metal.) Values for these coefficients, obtained for helicon propagation in a metal, were given in references [95] and [54], but a more general approach was presented in references [133] and [134], where spatial dispersion was taken into account. This approach made it possible to derive the transmission and reflection coefficients applicable for

DPR as well as for HPR, the latter being the limiting case of vanishing spatial dispersion. We will discuss them in detail immediately below, since the procedure for deriving these coefficients is common for both the Faraday effect in a plate and the Kerr effect.

## 4.6 The Kerr Effect

It was mentioned above that the acoustic analog of the Kerr effect manifests itself through the ellipticity and rotation of polarization that occur when ultrasound reflects from the interface between two media. We assume that the physical properties of one of the media depend upon the external magnetic field.

There are three variants of the Kerr effect that are distinguished by the direction of the static magnetic induction $\mathbf{B}_0$ with respect to the interfacial and incident planes: polar, for $\mathbf{B}_0 \parallel \mathbf{q}$; meridional (or longitudinal), for $\mathbf{q} \times \mathbf{k} \perp \mathbf{B}_0$; and equatorial (or transverse), for $\mathbf{q} \times \mathbf{k} \parallel \mathbf{B}_0$. Here $\mathbf{q}$ is the unit vector perpendicular to the interfacial plane. The linearly polarized incident waves are called $p$-type if the elastic displacement vector $\mathbf{u}$ is perpendicular to $\mathbf{q} \times \mathbf{k}$, and $s$-type if $\mathbf{u}$ is parallel to this vector.

Formally, this effect is the result of the boundary conditions that must be satisfied at the interface. Under these conditions one must account for the interaction of the elastic subsystem with the electromagnetic and magnetic subsystems and the conduction electrons. To provide a detailed description of the procedure for obtaining the reflection coefficients, in Sec. 4.6.1 we will discuss the most primitive case when the interaction is negligible and we can deal with the elastic subsystem alone. Later, in Secs. 4.6.2 and 4.6.3, the interactions between the subsystems will be taken into account.

### 4.6.1 Relations Among the Characteristic Parameters of Incident, Reflected, and Transmitted Elastic Waves

In discussing the classical problem of the reflection and transmission of a plane elastic wave, we will follow the approach given in [135]. Therefore, we will introduce a position vector in the interface, $\mathbf{r}_{int}$, a unit vector $\mathbf{q}$ perpendicular to the interface and directed from medium I to medium II ($\mathbf{r}_{int} \cdot \mathbf{q} = 0$), and a refraction vector $\mathbf{l}$ defined by

$$\mathbf{l} \cdot \mathbf{s} = \frac{\mathbf{k} \cdot \mathbf{s}}{\omega} = 1, \qquad (4.111)$$

where $\mathbf{s}$ is the phase velocity, as before. As noted above, the boundary conditions require continuity of the displacements at the interface (tight coupling of the media), and of the force acting on the interface (an absence of interfacial motion as a whole). With elastic displacements and tensions expressed as

$$\mathbf{u} = \mathbf{U}e^{i\varphi} = \mathbf{U}e^{-i\omega(\mathbf{l}\cdot\mathbf{r}_{int}-t)}, \tag{4.112}$$

$$\tau_{ij} = c_{ijkl}\frac{\partial u_l}{\partial x_k} = -i\omega c_{ijkl}l_k U_l e^{i\varphi} = \tau_{ij}^0 e^{i\varphi}, \tag{4.113}$$

the boundary conditions may be written as

$$\mathbf{U}^i e^{i\varphi^i} + \sum_{m=1}^{3} \mathbf{U}_m^r e^{i\varphi_m^r} - \sum_{n=1}^{3} \mathbf{U}_n^t e^{i\varphi_n^t} = 0, \tag{4.114}$$

$$i\omega^i c_{ijkl}^i l_k^i U_l^i e^{i\varphi^i} + \sum_{m=1}^{3} i\omega_m^r c_{ijkl}^r l_k^r U_l^r\, e^{i\varphi_m^r} - \sum_{n=1}^{3} i\omega_n^t c_{ijkl}^t l_k^t U_l^t e^{i\varphi_n^t} = 0. \tag{4.115}$$

The first terms in equations (4.114) and (4.115) correspond to the incident wave, symbols with superscript $r$ to the reflected waves, and with $t$ to the transmitted (refracted) ones; $c_{ijkl}^i = c_{ijkl}^r$ are the elastic moduli of the first medium and $c_{ijkl}^t$ are those of the second.

A linear combination of oscillatory functions can be equal to one another only if their arguments are equal. Hence, these equations are valid only if

$$\varphi^i = \varphi_m^r = \varphi_n^t. \tag{4.116}$$

Since this equality must hold for arbitrary $t$ and $\mathbf{r}$, it follows that

$$\omega^i = \omega_m^r = \omega_n^t \tag{4.117}$$

and

$$\mathbf{l}^i \cdot \mathbf{r}_{int} = \mathbf{l}_m^r \cdot \mathbf{r}_{int} = \mathbf{l}_n^t \cdot \mathbf{r}_{int},$$

or

$$\left(\mathbf{l}^i - \mathbf{l}_m^r\right) \cdot \mathbf{r}_{int} = \left(\mathbf{l}_m^r - \mathbf{l}_n^t\right) \cdot \mathbf{r}_{int} = \left(\mathbf{l}^i - \mathbf{l}_n^t\right) \cdot \mathbf{r}_{int} = 0. \tag{4.118}$$

As a consequence, all differences between any two refraction vectors should be parallel to $\mathbf{q}$ and perpendicular to $\mathbf{r}_{int}$, i.e., all of the refraction vectors belong to a common plane called the incident plane, which is defined by

$$(\mathbf{l}_l \times \mathbf{q}) \cdot \mathbf{r}_{int} = 0, \tag{4.119}$$

where $\mathbf{l}_l$ is any one of $\mathbf{l}^i$, $\mathbf{l}_m^r$ or $\mathbf{l}_n^t$. Also,

$$\mathbf{l}^i \times \mathbf{q} = \mathbf{l}_m^r \times \mathbf{q} = \mathbf{l}_n^t \times \mathbf{q} \equiv \mathbf{a}. \tag{4.120}$$

This equation gives the principal rules for reflection and refraction, namely,

$$\frac{\sin \theta^i}{s^i} = \frac{\sin \theta^r_m}{s^r_m} = \frac{\sin \theta^t_n}{s^t_n}, \tag{4.121}$$

where $\theta^i$ is the angle of incidence, $\theta^r_m$ that of reflection, corresponding to the wave with phase velocity $s^r_m$, and $\theta^t_n$ that of refraction, corresponding to the wave with phase velocity $s^t_n$. All of the angles are defined with respect to the normal $\mathbf{q}$ (see Fig. 4.10). By defining the vector $\mathbf{b}$ as

$$\mathbf{b} \equiv \mathbf{a} \times \mathbf{q} \tag{4.122}$$

and taking into consideration the definition (4.120), one can derive

$$\mathbf{l}_l = \xi_l \mathbf{q} + \mathbf{b}. \tag{4.123}$$

Here $\mathbf{l}_l$ has the same meaning as in equation (4.119) and $\xi_l$ is the $\mathbf{q}$ component of this vector. Note that, as is true for $\mathbf{a}$, vector $\mathbf{b}$ is the same for all the refraction vectors.

The universe of all refraction vectors corresponding to a particular mode may be represented as a surface in medium I for the reflected waves, or in II for the refracted ones. It is defined by

$$\mathbf{r} \equiv \mathbf{l}_l = \frac{\mathbf{n}_l}{s} \tag{4.124}$$

and called the inverse (velocity) or slowness surface [136] (here $\mathbf{n}_l$ is $\mathbf{k}_l/k_l$). The equation for this surface can be obtained using condition (4.8) for the existence of nontrivial solutions of the Christoffel equation (4.7).

By introducing the matrix $\mathbf{L}^n$ defined by

$$L^n_{ij} = \rho^{-1} c_{ijkl} n_k n_l, \tag{4.125}$$

we can write this condition in the form

$$\left| \mathbf{L}^n - s^2 \right| = 0, \tag{4.126}$$

which is equivalent to

$$s^6 - s^4 \mathrm{Sp}(\mathbf{L}^n) + s^2 \mathrm{Sp}(\overline{\mathbf{L}^n}) - |\mathbf{L}^n| = 0, \tag{4.127}$$

where Sp denotes the spur of a matrix and an overlined matrix is defined by $\overline{\mathbf{L}} \equiv |\mathbf{L}| \, \mathbf{L}^{-1}$.

If one takes into consideration the relations $s^2 \mathbf{L}^n = \mathbf{L}^s$, $s^4 \overline{\mathbf{L}^n} = \mathbf{L}^s$, and $s^6 |\mathbf{L}^n| = |\mathbf{L}^s|$, one can obtain the equation for the phase velocities,

$$s^{12} - s^8 \mathrm{Sp}(\mathbf{L}^s) + s^4 \mathrm{Sp}(\overline{\mathbf{L}^s}) - |\mathbf{L}^s| = 0. \tag{4.128}$$

Since three different velocities correspond to a particular direction $\mathbf{n}$, there are three inverse surfaces, two transverse and one longitudinal. Figure 4.10

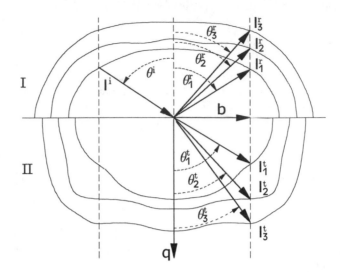

FIGURE 4.10. Incident ($\mathbf{l}^i$), reflected ($\mathbf{l}^r_m$), and refracted ($\mathbf{l}^t_n$) waves at the interface between two anisotropic media. The vector $\mathbf{b}$ lies in the interfacial plane and the vector $\mathbf{q}$ is perpendicular to it.

shows cross sections of the surfaces by the incident plane. If the media are isotropic, these surfaces degenerate into half spheres, two in each medium (because the transverse waves have a single velocity in an isotropic medium).

The intersections of the inverse surfaces and the plane defined by $\mathbf{r} = \mathbf{b}$ give the geometries of all refraction vectors. In order to obtain their analytical expressions, one must transform the Christoffel equation into a form written with respect to $\mathbf{l}_l$. Taking into account $\mathbf{L}^m = s^{-1}\mathbf{L}^n$, $\overline{\mathbf{L}}^m = s^{-4}\overline{\mathbf{L}}^n$, and $|\mathbf{L}^m| = s^{-6}|\mathbf{L}^n|$, the equation for the inverse surfaces may be written as

$$|\mathbf{L}^m - \mathbf{1}| = |\mathbf{L}^m| - \mathrm{Sp}(\overline{\mathbf{L}}^m) + \mathrm{Sp}(\mathbf{L}^m) - 1 = 0. \qquad (4.129)$$

We replace the vector $\mathbf{l}_l$ in this equation with $\xi_l\mathbf{q}$ and the known $\mathbf{b}$ to yield a sixth-order polynomial in $\xi$. If all the solutions are real, the three positive ones correspond to refracted waves and the three negative to reflected waves.

To determine the amplitudes of the reflected and refracted waves one must solve the Christoffel equation represented, for example, as

$$(L^m_{ij} - \delta_{ij})u_j = 0. \qquad (4.130)$$

In [135] it is shown that $u_i u_j = B(\overline{\mathbf{L}^m - \mathbf{1}})_{ij}$, so the elastic displacements may be written in the form

$$\mathbf{u} = A(\overline{\mathbf{L}^m - \mathbf{1}})\mathbf{p}, \tag{4.131}$$

where $\mathbf{p}$ is an arbitrary vector $(\mathbf{p} \cdot \mathbf{u} \neq 0)$.

After substituting expressions of the type (4.131) into the boundary conditions (4.114) and (4.115), we have

$$\mathbf{U}^i + \mathbf{U}_1^r + \mathbf{U}_2^r + \mathbf{U}_3^r - (\mathbf{U}_1^t + \mathbf{U}_2^t + \mathbf{U}_3^t) = 0 \tag{4.132}$$

and

$$\left\{ \left(\tau_{ij}^0\right)^i + \left(\tau_{ij}^0\right)_1^r + \left(\tau_{ij}^0\right)_2^r + \left(\tau_{ij}^0\right)_3^r - \left[\left(\tau_{ij}^0\right)_1^t + \left(\tau_{ij}^0\right)_2^t + \left(\tau_{ij}^0\right)_3^t\right] \right\} q_j$$

$$= 0, \tag{4.133}$$

where $\mathbf{U}^i$ and $\left(\tau_{ij}^0\right)^i$ are known characteristics of the incident wave, quasi-transverse or quasi-longitudinal.

Now we have six linear equations that contain six unknown coefficients defined by equation (4.131): $A_1^r$, $A_2^r$, $A_3^r$, $A_1^t$, $A_2^t$, and $A_3^t$. The solutions of the system (4.132)–(4.133) give us the amplitudes of the reflected and refracted waves. Finally, we write the transmission ($\mathbf{D}$) and reflection ($\mathbf{R}$) matrices as

$$\mathbf{U}_m^r = \mathbf{R}_{mn}\mathbf{U}_n^i, \qquad \mathbf{U}_m^t = \mathbf{D}_{mn}\mathbf{U}_n^i. \tag{4.134}$$

Here the index $m$ designates the type of the reflected or refracted wave, while $n$ designates the type of the incident wave.

### 4.6.2 Shear Elastic Waves Incident on the Interface between Dielectric and Ferromagnetic Media

In order to generalize the preceding subsection to deal with a magnetic dielectric medium, we must consider the magnetoelastic interaction discussed in Sec. 4.3, supplement the boundary conditions for elastic displacements and forces on the interface by those for magnetization, and take into account the contributions of all of the eigenmodes, including pure elastic as well as coupled waves.

The boundary conditions for magnetization were formulated in [53] as

$$\mathbf{m} = 0 \tag{4.135}$$

if the spins are fixed at the interface, and

$$\frac{\partial \mathbf{m}}{\partial z} = 0 \tag{4.136}$$

for free spins.

The other boundary conditions can be written in the form of equations (4.114)–(4.115). However, in general, the number of eigenmodes will be three (one longitudinal and two degenerate transverse) for the isotropic dielectric medium and five for the magnetic one. In addition, magnetoelastic tensions should be taken into account in equation (4.115).

The procedure for obtaining the dispersion equation for the waves in the magnetic medium is quite similar to the one described in Sec. 4.3.2, but now we will include energy dissipation in the spin system. This can be done by using the equation of motion for magnetization in the form of the Gilbert equation (4.55), since it takes into account the relaxation processes more accurately than the Landau–Lifshitz equation (4.54).

If the plane of incidence is given by $x = 0$, elastic displacements of the reflected shear waves $\mathbf{U}^r$ may be represented by a sum of mutually orthogonal vectors $\mathbf{U}_x^r$ and $\mathbf{U}_\perp^r$, where

$$\mathbf{U}_x^r = U_x^r \mathbf{e}_x \quad \text{and} \quad \mathbf{U}_\perp^r = U_\perp^r \mathbf{e}_\perp, \tag{4.137}$$

and $\mathbf{e}_x$ and $\mathbf{e}_\perp$ are unit vectors.

By introducing reflection coefficients $R_n^\pm$ for circular components defined by

$$R_n^\pm \equiv U_x^r / U_n^i \pm i U_\perp^r / U_n^i \tag{4.138}$$

and expressing them in the form $R_n^\pm = |R_n^\pm| \exp(i\rho_n^\pm)$, we obtain equations for the ellipticity and the rotation of the polarization of an elastic wave of $n$-type upon reflection from the interface between an isotropic dielectric medium and an anisotropic magnetic one as

$$\varepsilon = \frac{|R_n^+| - |R_n^-|}{|R_n^+| + |R_n^-|} \quad \text{and} \quad \phi = \frac{1}{2}\left(\rho_n^- - \rho_n^+\right). \tag{4.139}$$

Note that $n$ may refer to a shear mode of arbitrary polarization, as well as to an $s$ wave or a $p$ wave.

Clearly, the analysis of the principal features of the Kerr effect is too complicated in such a general approach, therefore we will consider a weak coupling case which reduces the order of the determinant and yields an analytical solution. Since we are assuming that medium II is a magnetic cubic uniaxial crystal with the easy axis parallel to [111], while $\mathbf{q}$ and $\mathbf{H}_0$ are along [001], we are dealing with the polar Kerr effect.

For this case the magnetoelastic field [i.e., the component of the effective field defined by equation (4.24)] is written as

$$(\mathbf{H}_{me})_k = -\frac{2}{M_0^2}\left[b_1 M_k \varepsilon_{kk} + b_2 M_k \varepsilon_{ki}\left(1 - \delta_{ki}\right)\right]. \tag{4.140}$$

Formally, we should use the superscript $t$ on all of the parameters and variables related to the magnetic medium, but we will defer this until such terms for all of the waves, incident, reflected, and transmitted, appear in the same expression.

In the linear approximation, with all variables assumed to be proportional to $\exp\left[i\left(\omega t - \mathbf{k} \cdot \mathbf{r}\right)\right]$,

$$\mathbf{H}_{me} = \frac{ib_2}{M_0}\left(k_1 u_3 + k_3 u_1,\; k_3 u_2 + k_2 u_3,\; 0\right). \tag{4.141}$$

The tensions for the properties of the magnetic medium have the form

$$\tau_{ij} = \frac{\partial W}{\partial \varepsilon_{ij}} = -\frac{i}{2}\left(c_{ijkl}k_k u_l + c_{ijkl}k_l u_k\right) + \frac{\partial W_{me}}{\partial \varepsilon_{ij}}. \tag{4.142}$$

Here the last term is found from the second equation of (4.37). Taking into account the relations between the nonvanishing elements for a cubic crystal, $c_{1111} = c_{2222} = c_{3333}$, $c_{1133} = c_{1122} = c_{2233}$, $c_{2323} = c_{1313} = c_{1212}$, we find

$$\tau_{13} = -i c_{1313} k_3 u_1 + \frac{b_2}{M_0}m_1,$$

$$\tau_{23} = -i c_{2323}(k_2 u_3 + k_3 u_2) + \frac{b_2}{M_0}m_2, \tag{4.143}$$

$$\tau_{33} = -i(c_{3322}k_2 u_2 + c_{3333}k_3 u_3) + b_1.$$

Since we are limiting ourselves to the case where all wave vectors lie in the $y - z$ plane, the complete system of equations has the form

$$a_{11}u_1 + a_{14}m_1 = 0,$$

$$a_{22}u_2 + a_{23}u_3 + a_{25}m_2 = 0,$$

$$a_{32}u_2 + a_{33}u_3 + a_{35}m_2 = 0, \tag{4.144}$$

$$a_{42}u_2 + a_{43}u_3 + a_{44}m_1 + a_{45}m_2 = 0,$$

$$a_{51}u_1 + a_{54}m_1 + a_{55}m_2 = 0,$$

where

$$a_{11} = \rho\omega^2 - c_{1313}(k_2^2 + k_3^2), \qquad a_{22} = \rho\omega^2 - c_{1111}k_2^2 - c_{1313}k_3^2,$$

$$a_{33} = \rho\omega^2 - c_{1111}k_3^2 - c_{1313}k_2^2, \quad a_{44} = a_{55} = i\omega,$$

$$a_{14} = a_{25} = -\frac{ib_2k_3}{M_0}, \qquad a_{23} = a_{32} = -(c_{1122} + c_{1313})k_2k_3,$$

$$a_{35} = -\frac{ib_2k_2}{M_0}, \qquad a_{42} = -a_{51} = i\gamma b_2 k_3,$$

$$a_{43} = i\gamma b_2 k_2, \qquad a_{54} = \gamma\left[\frac{2A}{M_0}(k_3^2 + k_2^2) + H_i\right] - i\alpha_0\omega,$$

$$a_{45} = -\gamma\left[\frac{2A}{M_0}(k_3^2 + k_2^2) + H_i + \frac{4\pi k_2^2 M_0}{k_2^2 + k_3^2}\right] + i\alpha_0\omega.$$

(4.145)

The rules for reflection and refraction (4.121) define $\mathrm{Re}(k_2)$ for all of the eigenvectors of a given frequency, so the only unknowns in the system (4.144) are $k_3$ and $\mathrm{Im}(k_2)$.

Using the last two equations of (4.144), we may express $m_1$ and $m_2$ as functions of $u_1$, $u_2$, and $u_3$, substitute these expressions into the first three equations, and set the determinant of the new system (containing these three equations) equal to zero:

$$(a_{11} + a_{14}b_{11})$$
$$\times\left[(a_{22} + a_{25}b_{22})(a_{33} + a_{35}b_{23}) - (a_{32} + a_{35}b_{22})(a_{23} + a_{25}b_{23})\right]$$
$$- a_{14}b_{12}\left[a_{25}b_{21}(a_{33} + a_{35}b_{23}) - a_{35}b_{21}(a_{23} + a_{25}b_{23})\right]$$
$$+ a_{14}b_{13}\left[a_{25}b_{21}(a_{32} + a_{35}b_{22}) - a_{35}b_{21}(a_{22} + a_{25}b_{22})\right] = 0,$$

(4.146)

where

$$b_{11} = \frac{a_{45}a_{51}}{a_{44}a_{55} - a_{54}a_{45}}, \quad b_{12} = \frac{-a_{55}a_{42}}{a_{44}a_{55} - a_{54}a_{45}}, \quad b_{13} = \frac{-a_{55}a_{43}}{a_{44}a_{55} - a_{54}a_{45}},$$

$$b_{21} = \frac{-a_{51}a_{44}}{a_{44}a_{55} - a_{54}a_{45}}, \quad b_{22} = \frac{a_{54}a_{42}}{a_{44}a_{55} - a_{54}a_{45}}, \quad b_{23} = \frac{a_{43}a_{54}}{a_{44}a_{55} - a_{54}a_{45}},$$

(4.147)

and $m_1$ and $m_2$ are defined by

$$m_i = b_{ij}u_j.$$

(4.148)

We will solve the dispersion equation (4.146) by an iterative method. The zero-order approximation will be obtained by letting the elements describing the magnon–phonon interaction (i.e., $a_{14}$, $a_{25}$, $a_{35}$, $a_{41}$, $a_{42}$, $a_{43}$,

$a_{51}$, $a_{52}$, and $a_{53}$) vanish. One of the solutions is

$$k_3^2 = \frac{\rho\omega^2}{c_{1313}} - k_2^2, \tag{4.149}$$

whereas the next two are

$$k_3^2 = \frac{-b \pm \left(b^2 - 4ac\right)^{1/2}}{2a}, \tag{4.150}$$

where

$$a = c_{1111}c_{1313},$$

$$b = c_{1111}^2 k_2^2 + c_{1313}^2 k_2^2 - \rho\omega^2 \left(c_{1313} + c_{1111}\right) - \left(c_{1122} + c_{1313}\right)^2 k_2^2,$$

$$c = \rho^2\omega^4 - \rho\omega^2 k_2^2 \left(c_{1313} + c_{1111}\right) + c_{1313}c_{1111}k_2^4, \tag{4.151}$$

Equations (4.149) and (4.150) for $k_3$ relate to pure elastic waves: one transverse $s$-type mode with wave vector $\mathbf{k}_s$, one quasi-transverse (quasi-$p$) mode with wave vector $\mathbf{k}_p$, and one quasi-longitudinal (quasi-$l$) mode with wave vector $\mathbf{k}_l$. At this point we have obtained real wave vectors (i.e., their imaginary parts are zero), since energy dissipation is considered to be in the magnetic subsystem only.

Next, the resulting expressions for $k_3$ should be substituted in turn into the equations which describe the magnetic subsystem and the magnon–phonon interaction. Recall that all $k_2$ in the zeroth approximation are real, equal, and known. Thus the solution for $k_3$ relating to the first weakly coupled phonon–magnon mode (now of the quasi-$s$ type, since it has $y$ and $z$ components of the elastic displacement) is

$$(k_3)_{qs}^2 = -k_2^2 + \frac{1}{c_{1313}}\Big\{\rho\omega^2 + a_{14}b_{11}$$
$$+ \left[(a_{22} + a_{25}b_{22})(a_{33} + a_{35}b_{23}) - (a_{32} + a_{35}b_{22})(a_{23} + a_{25}b_{23})\right]^{-1}$$
$$\times \big\{a_{14}b_{13}\left[a_{25}b_{21}(a_{32} + a_{35}b_{22}) - a_{35}b_{21}(a_{22} + a_{25}b_{22})\right]$$
$$- a_{14}b_{12}\left[a_{25}b_{21}(a_{33} + a_{35}b_{23}) - a_{35}b_{21}(a_{23} + a_{25}b_{23})\right]\big\}\Big\}, \tag{4.152}$$

where all $a_{ij}$ and $b_{mn}$ are functions of $(k_3)_s$.

The other two solutions relating to the coupled phonon-like (elastic-like) modes are written in the form of (4.150); however, the coefficient $c$ should now be given by

$$c = \rho^2\omega^4 - \rho\omega^2 \left[k_2^2 \left(c_{1313} + c_{1111}\right)\right] + c_{1313}c_{1111}k_2^4 + a_{35}a_{22}b_{23}$$
$$+ a_{25}a_{33}b_{22} - a_{32}a_{25}b_{23} - a_{35}a_{23}b_{22} - a_{14}\left[a_{11} + a_{14}b_{11}\right]^{-1}$$
$$\times \big\{b_{12}\left[a_{25}b_{21}(a_{33} + a_{35}b_{23}) - a_{35}b_{21}(a_{23} + a_{25}b_{23})\right]$$
$$- b_{13}\left[a_{25}b_{21}(a_{32} + a_{35}b_{22}) - a_{35}b_{21}(a_{22} + a_{25}b_{22})\right]\big\}. \tag{4.153}$$

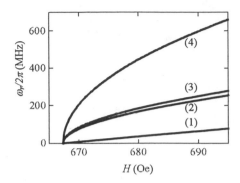

FIGURE 4.11. Magnetic field dependences of the resonance frequency for waves propagating in a YIG single crystal with **H** parallel to [001]. Curve (1) corresponds to a transverse wave propagating along **H**. Curves (2), (3), and (4) are for quasi-$s$, quasi-$p$, and quasi-$l$ modes, respectively, calculated for $\theta^i = 22.5°$.

Note that the solution of equation (4.150) with the $+$ sign before the root describes the coupled quasi-$p$ mode for which all $a_{ij}$ and $b_{kn}$ in (4.153) are functions of $(k_3)_p$, whereas that with the $-$ sign describes the coupled quasi-$l$ mode for which all $a_{ij}$ and $b_{kn}$ are functions of $(k_3)_l$.

Having obtained the complex $z$ components of the wave vectors, we can now determine the imaginary part of the $y$ components for a particular $\mathbf{k} = k\mathbf{e_k}$, where $k$ is a complex quantity and $\mathbf{e_k}$ is a unit vector in the $\mathbf{k}$ direction. Thus, the definition $\text{Re}(k_i) = \text{Re}[k \cos(\mathbf{k} \cdot \mathbf{e}_i)]$ should also be correct for the imaginary parts of the wave vector components. For every $\mathbf{k}_{qj}$ (where the index $j$ stands for $s, p,$ or $l$) it follows that

$$\text{Im}\,[(k_2)_{qj}] = \text{Im}\,[(k_3)_{qj}]\,\frac{\cos(\mathbf{k} \cdot \mathbf{e}_3)}{\cos(\mathbf{k} \cdot \mathbf{e}_2)} = \text{Im}\,[(k_3)_{qj}]\,\frac{\text{Re}\,[(k_2)_{qj}]}{\text{Re}\,[(k_3)_{qj}]}. \quad (4.154)$$

To illustrate the differences in the positions of magnon–phonon resonances for waves of different polarizations and to compare them with data for $\theta^i = 0$, Fig. 4.11 shows the resonance frequency $\omega_r$ versus magnetic field  calculated for a YIG single crystal. This frequency is determined by requiring that the real part of the resonant denominator $a_{44}a_{55} - a_{54}a_{45}$ in the definitions of $b_{ij}$ given by equations (4.147) vanish. Recall that $\theta^i$ is the angle of incidence, while the angle of propagation of the waves in the magnetic medium for $\theta^i \neq 0$ depends on $H$ at a particular frequency and therefore cannot be a fixed property of curves 2–4.

In the weak-coupling case now under consideration, the contribution of the magnon-like modes may be neglected, whereas complex amplitudes of the phonon-like modes should be used in boundary conditions of the type (4.132)–(4.133). These can be written by inserting their wave vectors into the system (4.144), and then the tensions will be given by equation (4.38).

In both cases $m_i$ is expressed in terms of $u_1$, $u_2$, and $u_3$ according to equation (4.148)).

Hence, the first three equations of (4.144), written in the form

$$
\begin{aligned}
c_{11}u_1 + c_{12}u_2 + c_{13}u_3 &= 0, \\
c_{21}u_1 + c_{22}u_2 + c_{23}u_3 &= 0, \\
c_{31}u_1 + c_{32}u_2 + c_{33}u_3 &= 0,
\end{aligned}
\tag{4.155}
$$

where

$$
\begin{aligned}
&c_{11} = a_{11} + a_{14}b_{11}, \quad c_{12} = a_{14}b_{12}, \quad\;\; c_{13} = a_{14}b_{13}, \\
&c_{21} = a_{25}b_{21}, \quad\qquad c_{22} = a_{22} + a_{25}b_{22}, \quad c_{23} = a_{23} + a_{25}b_{33}, \\
&c_{31} = a_{35}b_{21}, \quad\qquad c_{32} = a_{32} + a_{35}b_{22}, \quad c_{23} = a_{33} + a_{35}b_{33},
\end{aligned}
\tag{4.156}
$$

yield complex amplitudes expressed in terms of those which do not vanish for $b_2 \to 0$ and $\theta^i \to 0$ (i.e., in terms of $U_{1s}^t$, $U_{2p}^t$, and $U_{3l}^t$).

So in the magnetic medium we have the following expressions for the complex amplitudes of coupled phonon modes, where we denote their wave vectors by $\mathbf{k}_s^t$, $\mathbf{k}_p^t$, and $\mathbf{k}_l^t$, and omit the $q$ index:

(i) quasi-$s$ wave (all $a_{ij}$ and $b_{kl}$ are functions of $\mathbf{k}_s$):

$$
\begin{aligned}
U_{2s}^t &= \frac{c_{31}c_{23} - c_{21}c_{33}}{c_{22}c_{33} - c_{32}c_{23}} U_{1s}^t \equiv F_1(\mathbf{k}_s^t) U_{1s}^t \equiv A_{21}U_{1s}^t, \\
U_{3s}^t &= \frac{c_{32}c_{21} - c_{22}c_{31}}{c_{22}c_{33} - c_{32}c_{23}} U_{1s}^t \equiv F_2(\mathbf{k}_s^t) U_{1s}^t \equiv A_{31}U_{1s}^t,
\end{aligned}
\tag{4.157}
$$

(ii) quasi-$p$ wave:

$$
\begin{aligned}
U_{1p}^t &= \frac{c_{32}c_{13} - c_{12}c_{33}}{c_{11}c_{33} - c_{31}c_{13}} U_{2p}^t \equiv F_3(\mathbf{k}_p^t) U_{2p}^t \equiv A_{12}U_{2p}^t, \\
U_{3p}^t &= \frac{c_{31}c_{12} - c_{11}c_{32}}{c_{11}c_{33} - c_{31}c_{13}} U_{2p}^t \equiv F_4(\mathbf{k}_p^t) U_{2p}^t \equiv A_{32}U_{2p}^t,
\end{aligned}
\tag{4.158}
$$

(iii) quasi-$l$ wave:

$$
\begin{aligned}
U_{1l}^t &= \frac{c_{23}c_{12} - c_{13}c_{22}}{c_{11}c_{22} - c_{21}c_{12}} U_{2p}^t \equiv F_5(\mathbf{k}_l^t) U_{3l}^t \equiv A_{13}U_{3l}^t, \\
U_{2l}^t &= \frac{c_{21}c_{13} - c_{11}c_{23}}{c_{11}c_{22} - c_{21}c_{12}} U_{2p}^t \equiv F_6(\mathbf{k}_l^t) U_{3l}^t \equiv A_{23}U_{3l}^t.
\end{aligned}
\tag{4.159}
$$

In the nonmagnetic medium we have:

(i) an incident wave with (if we limit ourselves to a $p$ mode):

$$
\mathbf{k}_p^i = \frac{\omega}{s_p}(0, \sin\theta^i, \cos\theta^i), \quad \mathbf{U}_p^i = U_p^i(0, \cos\theta^i, -\sin\theta^i);
\tag{4.160}
$$

(ii) reflected waves with:

$$\mathbf{k}_s^r = \frac{\omega}{s_p}(0, \sin\theta^i, -\cos\theta^i), \qquad \mathbf{U}_s^r = U_s^r(1, 0, 0),$$

$$\mathbf{k}_p^r = \frac{\omega}{s_p}(0, \sin\theta^i, -\cos\theta^i), \qquad \mathbf{U}_p^r = U_p^r(0, \cos\theta^i, \sin\theta^i), \quad (4.161)$$

$$\mathbf{k}_l^r = \frac{\omega}{s_l}(0, \sin\theta_l^r, -\cos\theta_l^r), \qquad \mathbf{U}_l^r = U_l^r(0, -\sin\theta_l^r, \cos\theta_l^r),$$

where, according to equation (4.121),

$$\sin\theta_l^r = \frac{s_l}{s_p}\sin\theta^i, \qquad \cos\theta_l^r = \left[1 - \left(\frac{s_l}{s_p}\right)^2 \sin^2\theta^i\right]^{\frac{1}{2}}, \qquad (4.162)$$

and $U_s^r$ and $U_p^r$ are $U_x^r$ and $U_\perp^r$, respectively, in the notation of (4.137). The boundary conditions (4.132) can now be written as

$$U_s^r - U_{1s}^t - A_{12}U_{2p}^t - A_{13}U_{3l}^t = 0,$$

$$U_p^i\cos\theta^i + U_p^r\cos\theta^i - U_l^r\sin\theta_l^r - A_{21}U_{1s}^t - U_{2p}^t - A_{23}U_{3l}^t = 0, \quad (4.163)$$

$$-U_p^i\sin\theta^i + U_p^r\sin\theta^i + U_l^r\cos\theta_l^r - A_{31}U_{1s}^t - A_{32}U_{2p}^t - U_{3l}^t = 0.$$

In the boundary conditions for the tensions we should account for not only pure elastic but magnetoelastic components as well. Using the expressions (4.143) and, as before, writing the magnetization in terms of the elastic displacements as $m_k = b_{kj}u_j$, with the $b_{kj}$ given by (4.147), we can obtain the complex amplitudes of the tensions in the magnetic medium in the form

$$(\tau_{13}^0)_s^t = \left[-ic_{1313}(k_3)_s + \frac{b_2}{M_0}(b_{11} + b_{12}A_{21} + b_{13}A_{31})\right]U_{1s}^t \equiv (T_{13}^t)_s U_{1s}^t,$$

$$(\tau_{23}^0)_s^t = \left\{-ic_{2323}\left[(k_2)_s A_{31} + (k_3)_s A_{21}\right]\right.$$

$$\left. + \frac{b_2}{M_0}(b_{21} + b_{22}A_{21} + b_{23}A_{31})\right\}U_{1s}^t \equiv (T_{23}^t)_s U_{1s}^t,$$

$$(\tau_{33}^0)_s^t = -i\left[c_{3322}(k_2)_s A_{21} + c_{3333}(k_3)_s A_{31}\right]U_{1s}^t \equiv (T_{33}^t)_s U_{1s}^t,$$

$$(\tau_{13}^0)_p^t = \left[-ic_{1313}(k_3)_p A_{21} + \frac{b_2}{M_0}(b_{11}A_{12} + b_{12} + b_{13}A_{32})\right]U_{2p}^t$$

$$\equiv (T_{13}^t)_p U_{2p}^t,$$

$$(\tau_{23}^0)_p^t = \left\{-ic_{2323}\left[(k_2)_p A_{32} + (k_3)_p\right] + \frac{b_2}{M_0}(b_{21}A_{12} + b_{22} + b_{23}A_{32})\right\}U_{2p}^t$$

$$\equiv (T_{23}^t)_p U_{2p}^t,$$

$$(\tau_{33}^0)_p^t = -i\left[c_{3322}(k_2)_p + c_{3333}(k_3)_p A_{32}\right] U_{2p}^t \equiv (T_{33}^t)_p U_{2p}^t,$$

$$(\tau_{13}^0)_i^t = \left[-ic_{1313}(k_3)_l A_{13} + \frac{b_2}{M_0}(b_{11}A_{13} + b_{12}A_{23} + b_{13})\right] U_{3l}^t$$

$$\equiv (T_{13}^t)_l U_{3l}^t,$$

$$(\tau_{23}^0)_i^t = \left\{-ic_{2323}\left[(k_2)_l + (k_3)_l A_{23}\right] + \frac{b_2}{M_0}(b_{21}A_{13} + b_{22}A_{23} + b_{23})\right\} U_{3l}^t$$

$$\equiv (T_{23}^t)_l U_{3l}^t,$$

$$(\tau_{33}^0)_i^t = -i\left[c_{3322}(k_2)_l A_{23} + c_{3333}(k_3)_l\right] U_{3l}^t \equiv (T_{33}^t)_l U_{3l}^t. \tag{4.164}$$

Note that all of the coefficients $(T_{ab}^t)_j$ are functions of $\mathbf{k}_j$, where $j$ indicates the type of the wave (quasi-$s$, -$p$, or -$l$) in the magnetic medium.

The tensions in the nonmagnetic medium may be written in terms of the Lamé constants: $c_{1313} = c_{2323} = c_{1212} = \mu = \rho s_p^2 = \rho s_s^2$, $c_{1122} = c_{1133} = c_{3322} = \lambda$, and $c_{1111} = c_{2222} = c_{3333} = \lambda + 2\mu = \rho s_l^2$. Thus, the contributions of the incident wave are

$$(\tau_{13}^0)^i = 0, \qquad (\tau_{23}^0)^i = \left(-i\mu\frac{\omega}{s_p}\cos 2\theta^i\right) U_p^i \equiv T_{23}^i U_p^i,$$

$$(\tau_{33}^0)^i = \left(2i\mu\frac{\omega}{s_p}\sin 2\theta^i\right) U_p^i \equiv T_{33}^i U_p^i, \tag{4.165}$$

and those of the reflected ones are

$$(\tau_{13}^0)_s^r = \left(i\mu\frac{\omega}{s_p}\cos\theta^i\right) U_s^r \equiv (T_{13}^r)_s U_s^r,$$

$$(\tau_{23}^0)_p^r = \left(i\mu\frac{\omega}{s_p}\cos 2\theta^i\right) U_p^r \equiv (T_{23}^r)_p U_s^r,$$

$$(\tau_{33}^0)_p^r = \left(2i\mu\frac{\omega}{s_p}\sin 2\theta^i\right) U_p^r \equiv (T_{33}^r)_p U_p^r,$$

$$(\tau_{23}^0)_l^r = \left(-i\mu\frac{\omega}{s_l}\sin 2\theta_l^r\right) U_l^r \equiv (T_{23}^r)_l U_l^r,$$

$$(\tau_{33}^0)_l^r = \left[2i\frac{\omega}{s_l}(\lambda + 2\mu\cos^2\theta_l^r)\right] U_l^r \equiv (T_{33}^r)_l U_l^r,$$

$$(\tau_{23}^0)_s^r = (\tau_{33}^0)_s^r = (\tau_{13}^0)_p^r = (\tau_{13}^0)_l^r = 0. \tag{4.166}$$

Now we can express the boundary conditions (4.133) as

$$(T_{13}^r)_s U_s^r - (T_{13}^t)_s U_{3s}^t - (T_{13}^t)_p U_{3p}^t - (T_{13}^t)_l U_{3l}^t = 0,$$

$$T_{23}^i U_p^i + (T_{23}^r)_p U_p^r + (T_{23}^r)_l U_l^r - (T_{23}^t)_s U_{3s}^t - (T_{23}^t)_p U_{3p}^t - (T_{23}^t)_l U_{3l}^t = 0,$$

$$T_{33}^i U_p^i + (T_{33}^r)_p U_p^r + (T_{33}^r)_l U_l^r - (T_{33}^t)_s U_{3s}^t - (T_{33}^t)_p U_{3p}^t - (T_{33}^t)_l U_{3l}^t = 0. \tag{4.167}$$

To obtain the amplitudes of the reflected and refracted waves, it is more convenient to rewrite the systems of equations (4.163) and (4.167) in matrix form:

$$
\begin{pmatrix}
1 & A_{12} & A_{13} & -1 & 0 & 0 \\
A_{21} & 1 & A_{23} & 0 & -\cos\theta^i & \sin\theta^r_l \\
A_{31} & A_{32} & 1 & 0 & -\sin\theta^i & -\cos\theta^r_l \\
(T^t_{13})_s & (T^t_{13})_p & (T^t_{13})_l & -(T^r_{13})_s & 0 & 0 \\
(T^t_{23})_s & (T^t_{23})_p & (T^t_{23})_l & 0 & -(T^r_{23})_p & -(T^r_{23})_l \\
(T^t_{33})_s & (T^t_{33})_p & (T^t_{33})_l & 0 & -(T^r_{33})_p & -(T^r_{33})_l
\end{pmatrix}
\begin{pmatrix}
U^t_{1s}/U^i_p \\
U^t_{2p}/U^i_p \\
U^t_{3l}/U^i_p \\
U^r_s/U^i_p \\
U^r_p/U^i_p \\
U^r_l/U^i_p
\end{pmatrix}
$$

$$
=
\begin{pmatrix}
0 \\
\cos\theta^i \\
-\sin\theta^i \\
0 \\
T^i_{23} \\
T^i_{33}
\end{pmatrix}.
\qquad (4.168)
$$

Finally, we have a system of six inhomogeneous linear equations in six unknowns. Our particular interest is in $U^r_s/U^i_p$ and $U^r_p/U^i_p$, which yield the reflection coefficients

$$
R^\pm_p = U^r_s/U_p \pm iU^r_p/U^i_p. \qquad (4.169)
$$

Thus, after solving the system (4.168), we have obtained all the data required to determine $\varepsilon$ and $\phi$ from equations (4.139).

To illustrate the phenomena, Fig. 4.12(b) shows the Kerr rotation and ellipticity that occur at a quartz–YIG interface. Calculations performed on the basis of the scheme given above show that variations of the wave vector related to a quasi-$s$ mode exceed those of quasi-$p$ and quasi-$l$ modes by factors of $10^2$ and $10^3$, respectively. As a consequence, only the $z$ component of $\Delta\mathbf{k}_s$ is shown in Fig. 4.12(a) to compare with the behaviors of $\varepsilon(H)$ and $\phi(H)$. In contrast to the case of Faraday rotation and ellipticity in the weak-coupling regime, which are proportional to the real and imaginary parts of $\Delta\mathbf{k}$ for the resonant wave, here $\phi(H)$ and $\varepsilon(H)$ are proportional to the imaginary and real parts, respectively.

### 4.6.3   Shear Elastic Waves Incident on the Interface between Dielectric and Metallic Media

Next we come to the problem of reflection and transmission of elastic waves at the interface between a dielectric and a non-ferromagnetic metal. In contrast with the previous case, the interaction of the elastic subsystem with the conduction electrons should be properly accounted for. It results in the excitation of coupled modes by the incident linearly polarized elastic wave, and reflection of both electromagnetic and elastic modes with polarization

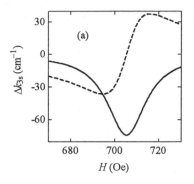 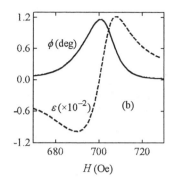

FIGURE 4.12. (a) Magnetic field dependences of the real (solid line) and imaginary (dashed line) parts of the $z$ component of the wave vector of a quasi-$s$ wave with a frequency of $300\,\mathrm{MHz}$ propagating in a YIG single crystal for an angle of incidence at the quartz interface of $\theta^i = 22.5°$ calculated with $\alpha_0 = 0.1$; $\Delta\mathbf{k}_s \equiv \mathbf{k}_s(H) - \mathbf{k}_s(0)$. (b) Magnetic field dependences of the rotation of the polarization (solid line) and the ellipticity (dashed line).

differing from the polarization of the incident wave. By taking it into account we will obtain all of the equations necessary for a description of the Kerr and Faraday effects in a metal plate. The presentation will be based on the contemporary approach to the problem with spatial dispersion taken into account [133]. This method will give us the opportunity to provide a general treatment of both DPR and HPR.

Basic equations

Let us consider normal incidence of a shear, linearly polarized elastic wave, represented by $\mathbf{k}_u^i$ in Fig. 4.13, on a plane interface ($z = 0$) between a dielectric medium (I) and a metal (II). The reflected elastic and electromagnetic

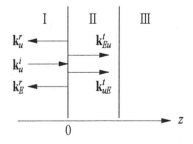

FIGURE 4.13. Reflected and transmitted elastic and electromagnetic waves for an elastic plane wave normally incident on a plane interface between media I and II.

modes are represented by $\mathbf{k}_u^r$ and $\mathbf{k}_E^r$, respectively, and the transmitted coupled modes by $\mathbf{k}_{Eu}^t$ and $\mathbf{k}_{uE}^t$. The third medium (III) is assumed to be identical to medium I and the second interface is also a plane, parallel to the first one. The external magnetic field $\mathbf{H}$ as well as all wave vectors are collinear to the $z$ axis, while the variable electric $\mathbf{E}$ and magnetic $\mathbf{h}$ fields, as well as the elastic displacement, have only transverse components with respect to the wave vectors for this geometry.

To determine $\mathbf{E}$, $\mathbf{h}$, and $\mathbf{u}$ in the metal, one should solve the system of equations (4.75)–(4.76), but, rather than assume infinite media, supplement them with the appropriate boundary conditions at $z = 0$:

$$\mathbf{E}_t = \mathbf{E}_i + \mathbf{E}_r, \qquad\qquad \mathbf{h}_t = \mathbf{h}_i + \mathbf{h}_r,$$

$$\mathbf{u}_t = \mathbf{u}_i + \mathbf{u}_r, \qquad (\tau_{kl}^{tot})_t n_l = (\tau_{kl}^{tot})_i n_l + (\tau_{kl}^{tot})_r n_l, \qquad (4.170)$$

where, as before, subscripts $i$, $r$, and $t$ relate to the incident, reflected, and transmitted waves, respectively, $\mathbf{n}$ is a unit vector normal to the interface, and

$$\tau_{kl}^{tot} = c_{klmn} \varepsilon_{mn} + \tau_{kl}^d + \tau_{kl}^M \qquad (4.171)$$

is the component of the total tension tensor involving the quasi-static elastic, the dynamic [defined by equation (4.71)], and the Maxwell contributions,

$$\tau_{kl}^M = \frac{1}{4\pi} \left\{ E_k E_l + (\mathbf{h} + \mathbf{H})_k (\mathbf{h} + \mathbf{H})_l - \frac{\delta_{kl}}{2} \left[ E^2 + (\mathbf{h} + \mathbf{H})^2 \right] \right\},$$

$$\frac{\partial \tau_{kl}^M}{\partial x_l} = \frac{1}{c} (\mathbf{j} \times \mathbf{H})_k. \qquad (4.172)$$

The boundary conditions (4.170) should be supplemented by those for the distribution function given by equation (4.60). With the assumption of specular reflection of electrons from the interface, it is

$$\chi(z, v_z)|_{z=0} = \chi(z, -v_z)|_{z=0}. \qquad (4.173)$$

In order to make further progress in solving this problem we must make some simplifications. In particular, the FS will be assumed to have axial symmetry and the deformation potential to be a function of $v_z$. By introducing the even and odd components of $\chi$ with respect to $v_z$, as was done in [137] and [138], extending the variable fields and displacements out of the metal as even functions of $z$, and carrying out a cosine-Fourier transform, one obtains the following analogs of the earlier material equations (4.70)–(4.71) expressed in terms of Fourier components:

$$j_m(k) = \sigma_{mn}(k) F_n(k) + i\omega p_{m,zl}(k) u_l(k), \qquad (4.174)$$

$$\frac{\partial \tau_{mn}^d(k)}{\partial x_l} = p_{mz,n}(k)F_l(k) - i\omega S_{mzlz}(k)u_l(k), \qquad (4.175)$$

where

$$\sigma_{mn}(k) = -\frac{2e^2}{\Omega h^3} \int d^3p\, v_m \frac{\partial f_0}{\partial \varepsilon} \int_{-\infty}^{\varphi} d\varphi'\, v_n\,(\varphi') \cos\left[k\frac{v_z}{\Omega}(\varphi - \varphi')\right]$$
$$\times \exp\left[i\left(\frac{\omega}{\Omega} - i\gamma\right)(\varphi' - \varphi)\right], \qquad (4.176)$$

$$p_{m,zl}(k) = \frac{2ek}{\Omega h^3} \int d^3p\, v_m \frac{\partial f_0}{\partial \varepsilon} \int_{-\infty}^{\varphi} d\varphi'\, \Lambda_{lz}\,(\varphi') \sin\left[k\frac{v_z}{\Omega}(\varphi - \varphi')\right]$$
$$\times \exp\left[i\left(\frac{\omega}{\Omega} - i\gamma\right)(\varphi' - \varphi)\right], \qquad (4.177)$$

$$p_{mz,l}(k) = \frac{2ek}{\Omega h^3} \int d^3p\, \Lambda_{mz} \frac{\partial f_0}{\partial \varepsilon} \int_{-\infty}^{\varphi} d\varphi'\, v_l\,(\varphi') \sin\left[k\frac{v_z}{\Omega}(\varphi - \varphi')\right]$$
$$\times \exp\left[i\left(\frac{\omega}{\Omega} - i\gamma\right)(\varphi' - \varphi)\right], \qquad (4.178)$$

$$S_{mzlz}(k) = -\frac{2k^2}{\Omega h^3} \int d^3p\, \Lambda_{mz} \frac{\partial f_0}{\partial \varepsilon} \int_{-\infty}^{\varphi} d\varphi'\, \Lambda_{lz}\,(\varphi') \cos\left[k\frac{v_z}{\Omega}(\varphi - \varphi')\right]$$
$$\times \exp\left[i\left(\frac{\omega}{\Omega} - i\gamma\right)(\varphi' - \varphi)\right]. \qquad (4.179)$$

If one compares these equations with (4.70) and (4.71), one can see that $(-ik)^{-1}p_{mnl}$ and $-k^{-2}S_{mnlk}$ play the roles of the tensor components $\beta_{mnl}$ and $\alpha_{mnlk}$, respectively, defined for an infinite metal.

Thus, Maxwell's equations and the equations of elasticity theory written for Fourier components give

$$k^2 E_m(k) + 2E_m'(0) = \frac{4\pi i\omega}{c^2} j_m(k) \qquad (4.180)$$

and

$$-\rho\omega^2 u_m(k) + c_{mzzn}\delta_{mn}\left[k^2 u_n(k) + 2u_n'(0)\right]$$
$$= \frac{1}{c}\left[\mathbf{j}(k) \times \mathbf{H}\right]_m + p_{mz,n}(k)F_n(k) - i\omega S_{mznz}u_n(k), \qquad (4.181)$$

where $m, n = x, y$, $j_z(k) = 0$, and

$$E_m'(0) \equiv \left.\frac{\partial E_m}{\partial z}\right|_{z=0}, \qquad u_n'(0) \equiv \left.\frac{\partial u_n}{\partial z}\right|_{z=0}.$$

Recall that $\rho$ and $c_{mzzn}$ are parameters of the metal (i.e., medium II). By introducing circular components [(4.80) and (4.81)] as before, solving the system of equations (4.180)–(4.181), and performing the inverse Fourier transform, one obtains equations for the displacement and the electric field in the metal:

$$u^\pm(z) = u'_\pm(0) J^\pm_{uu}(z) + E'_\pm(0) J^\pm_{uE}(z), \tag{4.182}$$

$$E^\pm(z) = u'_\pm(0) J^\pm_{Eu}(z) + E'_\pm(0) J^\pm_{EE}(z), \tag{4.183}$$

where

$$J^\pm_{uu}(z) = -\frac{2}{\pi} \int_0^\infty dk \frac{\cos kz}{\Delta_\pm} \left( k^2 + \frac{4\pi i \omega \sigma^\pm}{c^2} \right), \tag{4.184}$$

$$J^\pm_{uE}(z) = \mp \frac{2}{\pi} \frac{\eta c}{\omega H} \int_0^\infty dk \frac{\cos kz}{\Delta_\pm} \frac{4\pi i \omega \sigma^\pm}{c^2} \left( 1 \mp \frac{icp^\pm}{H\sigma^\pm} \right), \tag{4.185}$$

$$J^\pm_{Eu}(z) = \mp \frac{2}{\pi} \frac{\omega H}{c} \int_0^\infty dk \frac{\cos kz}{\Delta_\pm} \frac{4\pi i \omega \sigma^\pm}{c^2} \left( 1 \mp \frac{icp^\pm}{H\sigma^\pm} \right), \tag{4.186}$$

$$J^\pm_{EE}(z) = -\frac{2}{\pi} \int_0^\infty dk \frac{\cos kz}{\Delta_\pm}$$
$$\times \left[ k^2 - \frac{\rho \omega^2}{c_{1313}} + \frac{i\omega S^\pm}{c_{1313}} + \eta \frac{4\pi i \omega \sigma^\pm}{c^2} \left( 1 \mp \frac{2icp^\pm}{H\sigma^\pm} \right) \right], \tag{4.187}$$

and $\eta$ is the coupling parameter defined by equation (4.88). The equation

$$\Delta_\pm \equiv \left( k^2 + \frac{4\pi i \omega \sigma^\pm}{c^2} \right) \left[ k^2 - \frac{\rho \omega^2}{c_{1313}} + \frac{i\omega S^\pm}{c_{1313}} + \eta \frac{4\pi i \omega \sigma^\pm}{c^2} \left( 1 \mp \frac{2icp^\pm}{H\sigma^\pm} \right) \right]$$
$$- \eta \left[ \frac{4\pi i \omega \sigma^\pm}{c^2} \left( 1 \mp \frac{icp^\pm}{H\sigma^\pm} \right) \right]^2 = 0, \tag{4.188}$$

determines the wave numbers of the circularly polarized normal modes in the metal.

As a result of the incidence of the shear elastic wave

$$u_i^\pm(z, t) = U_i^\pm \exp\left[ i\left( \omega t - k_0 z \right) \right] \tag{4.189}$$

on the interface at $z = 0$, elastic and electromagnetic reflected waves will propagate in the dielectric (medium I):

$$u_r^\pm(z, t) = U_r^\pm \exp\left[ i\left( \omega t + k_0 z \right) \right],$$

$$E_r^\pm(z, t) = E_r^\pm \exp\left[ i\left( \omega t + k_0^E z \right) \right]. \tag{4.190}$$

To obtain the elastic wave reflection coefficients $R^\pm$ defined by equations (4.134), one should substitute equation (4.182) into the third equation of (4.170). The $\tau_\pm^M$ are identical for media I and II, whereas $\tau_\pm^d|_{z=0} = 0$ since the $x$ and $y$ components of electron velocity do not change under specular reflection from the interface. Recall that just this restriction on the distribution function (4.173) has been assumed in our treatment. Thus we have

$$\rho_0 s_0^2 \left. \left( \frac{\partial u_i^{\pm}}{\partial z} + \frac{\partial u_r^{\pm}}{\partial z} \right) \right|_{z=0} = \left. \rho s^2 \frac{\partial u^{\pm}}{\partial z} \right|_{z=0} \tag{4.191}$$

and

$$R^{\pm} = \frac{\rho_0 s_0 - \rho s \left[ -i k_s J_{uu}^{\pm}(0) \right]^{-1}}{\rho_0 s_0 + \rho s \left[ -i k_s J_{uu}^{\pm}(0) \right]^{-1}} \equiv \frac{Z_0 - Z^{\pm}}{Z_0 + Z^{\pm}}, \tag{4.192}$$

where $s$ is the quasi-static phase velocity of the elastic wave in medium II (found from $\rho s^2 \equiv c_{1313}$), $k_s = \omega / s$, and the elastic displacements in the metal, $u^{\pm}(z)$, are defined by equation (4.182). The authors of [133] state that the contribution of $J_{uE}^{\pm}(0)$ may be neglected here.

If we define $\rho^{\pm}$ by $R^{\pm} = |R^{\pm}| \exp(i\rho^{\pm})$, we may write the equations for the ellipticity and the rotation angle of the polarization of the initially linearly polarized elastic shear wave caused by reflection from the dielectric–metal interface in the same way as those given in equation (4.139), namely,

$$\varepsilon = \frac{|R^+| - |R^-|}{|R^+| + |R^-|}, \qquad \phi = \frac{1}{2} \left( \rho^- - \rho^+ \right). \tag{4.193}$$

Helicon–phonon resonance

Let us consider the limiting case relating to the absence of spatial dispersion. It corresponds to the vanishing of the electroacoustic coefficients $\beta^{\pm}$ and (usually) $\alpha^{\pm}$ when $\sigma^{\pm}$ is independent of the FS shape. Electromagnetic waves in a noncompensated metal occur as weakly damped helicons in one of the circular polarizations and strongly damped waves in the other. The dispersion of these waves is given by equation (4.83), while the dispersion of the coupled waves has the form given in equation (4.87). Substituting the left-hand side of the latter equation into (4.184) instead of $\Delta_{\pm}$, we find

$$-i k_s J_{uu}^{\pm}(0) = \left( k_s + k_E^{\pm} \right) \left[ \left( k_s + k_E^{\pm} \right)^2 + \eta \left( k_E^{\pm} \right)^2 \right]^{-\frac{1}{2}}, \tag{4.194}$$

and the reflection coefficients are given by

$$R^{\pm} = \frac{1 - \frac{\rho s}{\rho_0 s_0} \left[ 1 + \eta \frac{\left( k_E^{\pm} \right)^2}{\left( k_s + k_E^{\pm} \right)^2} \right]^{\frac{1}{2}}}{1 + \frac{\rho s}{\rho_0 s_0} \left[ 1 + \eta \frac{\left( k_E^{\pm} \right)^2}{\left( k_s + k_E^{\pm} \right)^2} \right]^{\frac{1}{2}}}. \tag{4.195}$$

Owing to the smallness of $\eta$ and $\gamma$, we may neglect terms proportional to $\eta\gamma$ and obtain the following equations for a metal in which the helicon has

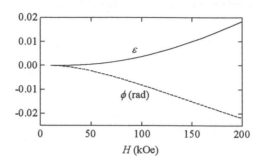

FIGURE 4.14. Ellipticity (solid curve) and rotation of the polarization (dashed curve) of ultrasound as functions of magnetic field for coupled helicon–phonon modes for reflection from the interface between a dielectric and a metal. The properties of the dielectric medium ($\rho_0 = 2.2\,\mathrm{g/cm^3}$, $s_0 = 3.8 \times 10^5$ cm/s) correspond to amorphous quartz; the other parameters are the same as for Fig. 4.4.

($-$) polarization:

$$\phi = \frac{1 - R_0^2}{4R_0} \frac{k_0/k_E}{\left[1 + (k_0/k_E)^2\right]^2} \eta, \tag{4.196}$$

$$\varepsilon = \frac{1 - R_0^2}{8R_0} \left\{ \frac{(k_0/k_E)^2 - 1}{\left[1 + (k_0/k_E)^2\right]^2} + \frac{1}{(1 + k_0/k_E)^2} \right\} \eta, \tag{4.197}$$

where

$$R_0 = \frac{\rho_0 s_0 - \rho s}{\rho_0 s_0 + \rho s}, \quad k_E = \left(4\pi\omega\sigma_{21}/c^2\right)^{1/2} \tag{4.198}$$

are the reflection coefficient calculated without accounting for the conduction electrons, and the real component of $k_E^-$, respectively.

Equations (4.196) and (4.197) were presented in [53]. It was noted (and Fig. 4.14 confirms the statement) that the polarization parameters do not have a resonant character. Moreover, the rotation of the polarization is determined by the interaction of the elastic wave with the nonpropagating electromagnetic wave which was observed for the polarization opposite to the polarization of the helicon.

Doppleron–phonon resonance

When discussing the effects initiated by spatial dispersion, the following points should be mentioned. The spectrum of the electromagnetic waves (dopplerons) depends strongly on the form of the FS. As a result of a complicated dependence of conductivity on $k$, there may be several roots of the dispersion equation relating to weakly damped circularly polarized modes. Some of them correspond to waves having phase and group velocities in parallel directions, but a more typical situation is that where energy propagation and the wave vector are in opposite directions. In the latter case, if the coupling parameter is large and the losses are small, a gap appears in the spectrum of the coupled doppleron–phonon modes; some interval of $\omega$ becomes forbidden for wave excitation. This feature leads to the appearance of resonances in the dependences of $\phi$ and $\varepsilon$ on $\omega$ or $H$.

Since it is impossible to analyze the expressions for the polarization parameters in a general form without models for the FS and deformation potential and, furthermore, a discussion of particular cases would occupy too much space, we will not consider the Kerr effect with DPR here. Anyone interested in this topic should read Reference [133].

## 4.7   The Faraday Effect in a Plate

Another experimental situation that enables one to observe the Faraday effect is wave propagation in a plate sandwiched between two isotropic media, as shown in Fig. 4.13. It was described in [134] in the form presented here.

Again, normal incidence will be discussed and, as before, we will consider media I and III to be identical dielectrics while medium II is a metal. The locations of the plane interfaces will correspond to $z = 0$ and $z = z_0$, with wave propagation along the positive direction of the $z$ axis. We will denote values of variables at the first and second interfaces with the subscripts 1 and 2, respectively. Assuming that the pulse technique will be used, our description of the polarization effects will refer to experiments employing a metal plate with $z_0 \gg \lambda_s$, where $\lambda_s$ is the wavelength of the elastic wave. The pulse duration for this case (usually of the order of $1\,\mu s$) is less than the transit time of sound through the specimen, so interference of the multiple reflections from the interfaces can be neglected here.

Variable elastic and electromagnetic fields in the metal will be written in the form given in equations (4.182) and (4.183), with $J_{ab1}^{\pm}$ substituted for $J_{ab}^{\pm}$:

$$u^{\pm}(z) = u'_{\pm}(0)J_{uu1}^{\pm}(z) + E'_{\pm}(0)J_{uE1}^{\pm}(z), \tag{4.199}$$

$$E^{\pm}(z) = u'_{\pm}(0)J_{Eu1}^{\pm}(z) + E'_{\pm}(0)J_{EE1}^{\pm}(z). \tag{4.200}$$

The boundary conditions, given by equations (4.170), along with the small size of the electromagnetic wave number in the dielectric compared with the wave numbers of the elastic and electromagnetic waves in the metal, lead to

$$E^{\pm\prime}(0) = 0, \tag{4.201}$$

$$u^{\pm\prime}(0) = \frac{-2ik_0\left(\rho_0 s_0^2/\rho s^2\right)U_i^{\pm}}{1 - ik_0\left(\rho_0 s_0^2/\rho s^2\right)J_{uu1}^{\pm}(0)}, \tag{4.202}$$

where $U_i^{\pm}$ is defined in equation (4.189).

The waves reflected at the second interface may be written as

$$u_r^{\pm}(z) = -u_{\pm}'(z_0)J_{uu2}^{\pm}(z_0 - z) - E_{\pm}'(z_0)J_{uE2}^{\pm}(z_0 - z), \tag{4.203}$$

$$E_r^{\pm}(z) = -u_{\pm}'(z_0)J_{Eu2}^{\pm}(z_0 - z) - E_{\pm}'(z_0)J_{EE2}^{\pm}(z_0 - z), \tag{4.204}$$

where $E_{\pm}'(z_0)$ and $u_{\pm}'(z_0)$ are derivatives with respect to $z$ at $z = z_0$. In medium III the elastic waves may be represented by

$$u_3^{\pm}(z,t) = U_3^{\pm}\exp\left[i\left(\omega t - k_0 z\right)\right]. \tag{4.205}$$

Using the boundary conditions at $z = z_0$ and taking into account the small value of the electromagnetic wave number in a dielectric, one can obtain the amplitude transmission coefficients for the elastic displacement:

$$D^{\pm} \equiv \frac{U_3^{\pm}}{U_i^{\pm}} = -2ik_0\left(\rho_0 s_0^2/\rho s^2\right)$$

$$\times \frac{J_{uu1}^{\pm}(z_0) + J_{uu2}^{\pm}(0)J_{uu1}^{\pm\prime}(z_0) + J_{uE2}^{\pm}(0)J_{Eu1}^{\pm\prime}(z_0)}{\left[1 - ik_0\left(\rho_0 s_0^2/\rho s^2\right)J_{uu1}^{\pm}(0)\right]\left[1 - ik_0\left(\rho_0 s_0^2/\rho s^2\right)J_{uu2}^{\pm}(0)\right]}, \tag{4.206}$$

where $J_{ab}^{\pm\prime}(z_0) = \left.\dfrac{\partial J_{ab}^{\pm}}{\partial z}\right|_{z=z_0}$.

By writing the transmission coefficients in a form similar to that of the reflection coefficients, namely, $D^{\pm} = |D^{\pm}|\exp(i\delta^{\pm})$, we can write the equations for the polarization parameters as

$$\varepsilon = \frac{|D^+| - |D^-|}{|D^+| + |D^-|}, \qquad \phi = \frac{1}{2}\left(\delta^- + \delta^+\right). \tag{4.207}$$

Helicon–phonon resonance

First, let us consider the case of no spatial dispersion, starting with HPR under weak coupling, i.e., $\eta \ll \gamma^2$. The solutions of the dispersion equation (4.87) for the helicon-like and phonon-like modes are

$$k_{hl}^{\pm\,2} = k_E^{\pm\,2}\left(1 - \eta\frac{k_E^{\pm\,2}}{k_s^2 - k_E^{\pm\,2}}\right), \tag{4.208}$$

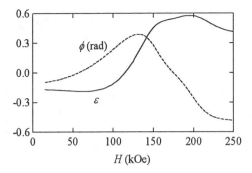

FIGURE 4.15. Calculated ellipticity (solid curve) and rotation of the polarization (dashed curve) of ultrasound as functions of magnetic field after passing through a metal plate of thickness $z_0 = 0.8\,\mathrm{mm}$ between plates of amorphous quartz for coupled helicon–phonon modes. All of the parameters are the same as for Fig. 4.14, except that $\tau = 10^{-12}\,\mathrm{s}$, which corresponds to the weak-coupling case.

$$k_{pl}^{\pm 2} = k_s^2 \left( 1 + \eta \frac{k_E^{\pm 2}}{k_s^2 - k_E^{\pm 2}} \right). \tag{4.209}$$

By neglecting the nonresonant contributions to the transmission coefficients that are due to the presence of a surface, one can obtain the following expressions for a metal in which the weakly damped helicon has $(+)$ polarization:

$$D^+ = \frac{4\left(\rho_0 s_0/\rho s\right)}{\left[1 + \left(\rho_0 s_0/\rho s\right)\right]^2} \left\{ \left[ 1 - \eta \frac{k_s^4}{\left[k_s^2 - k_E^2\left(1 + i\gamma\right)\right]^2} \right] \exp\left(-ik_{pl}^+ z_0\right) \right.$$

$$\left. + \eta \frac{k_s^4}{\left[k_s^2 - k_E^2\left(1 + i\gamma\right)\right]^2} \exp\left(-ik_{hl}^+ z_0\right) \right\}$$

$$\equiv D_{pl}^+(0) \exp\left(-ik_{pl}^+ z_0\right) + D_{hl}^+(0) \exp\left(-ik_{hl}^+ z_0\right), \tag{4.210}$$

$$D^- = \frac{4\left(\rho_0 s_0/\rho s\right)}{\left[1 + \left(\rho_0 s_0/\rho s\right)\right]^2} \exp\left(-ik_s z_0\right), \tag{4.211}$$

where $D_{pl}^+(0)$ and $D_{hl}^+(0)$ are the transmission coefficients, independent of propagation path length in the metal, for the phonon- and helicon-like modes, respectively. Recall that $k_E$ was defined in equation (4.198).

Figure 4.15 shows $\varepsilon(H)$ and $\phi(H)$ calculated from equations (4.207)–(4.211) using the parameters of indium for the metal and amorphous quartz for the dielectric. It should be noted that the shapes of these curves are quite different from typical ones for bulk polarization phenomena under weak coupling. The latter are similar to the dependences of the wave number and absorption of the phonon-like mode on the field, whereas the curves

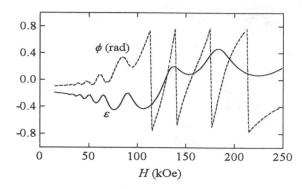

FIGURE 4.16. Calculated ellipticity (solid curve) and rotation of the polarization (dashed curve) of ultrasound as functions of magnetic field after passage through a metal plate of thickness $z_0 = 0.8\,\mathrm{mm}$ between plates of amorphous quartz for coupled helicon–phonon modes. All of the parameters are the same as for Fig. 4.15, except $\tau = 10^{-11}\,$s, which corresponds to the strong coupling case.

given in Fig. 4.15 exhibit distortions caused by the complex nature of the transmission coefficients $D_{pl}^+(0)$ and $D_{hl}^+(0)$.

In discussing the case of strong coupling ($\eta \gg \gamma^2$), $D^+$ should be written as

$$D^+ = \frac{2\rho_0 s_0}{\rho s} \frac{D_1^+(0)\exp\left(-ik_1^+ z_0\right) - D_2^+(0)\exp\left(-ik_2^+ z_0\right)}{\left(k_1^+ + k_2^+\right)\left(k_1^{+2} - k_2^{+2}\right)\left[1 + \frac{\rho_0 s_0}{\rho s}\frac{k_s + k_E^+}{k_1^+ + k_2^+}\right]^2}, \quad (4.212)$$

where $k_1^+$ and $k_2^+$ are the wave numbers of the coupled helicon–phonon modes, defined by equation (4.89), while $D_{1,2}^+(0)$ may be written as

$$D_1^+(0) = \left(k_1^+ k_s - k_2^+ k_E^+\right)\left(k_1^+ + k_2^+\right) + \left(k_s + k_E^+\right)\left(k_1^{+2} - k_E^{+2}\right),$$

$$D_2^+(0) = \left(k_2^+ k_s - k_1^+ k_E^+\right)\left(k_1^+ + k_2^+\right) + \left(k_s + k_E^+\right)\left(k_2^{+2} - k_E^{+2}\right). \quad (4.213)$$

As before, the transmission coefficient for the wave of nonresonant polarization $D^+$ may be regarded as one corresponding to the quasi-static regime [see equation (4.211)]. Typical curves for ellipticity and rotation of the polarization under strong coupling are given in Fig. 4.16. The oscillatory character of the dependences is the product of the interference between the two coupled modes which exist for resonant polarization. Discontinuities of $\phi(H)$ are present because only the principal values of $\arctan\left(\mathrm{Im}\,D^\pm/\mathrm{Re}\,D^\pm\right)$, which define $\delta^\pm$ in equation (4.207), are given.

# 5
# Experiments in Magnetic Materials

## 5.1 The Faraday Effect

In 1960 Bömmel and Dransfeld reported the first observation of rotation of polarization in a ferromagnet [8]. They investigated the propagation of shear ultrasound through a thin disk of YIG magnetized in the hard direction near ferromagnetic resonance. In addition to strong absorption, the authors discovered that the polarization rotated about 45° in a disk only 0.010 inch (0.25 mm) thick at $f = 300\,\mathrm{MHz}$ and room temperature. They did not mention the orientation of the wave vector with respect to the **B** but, since they only reported rotation of the polarization, we conclude that the disk was magnetized perpendicularly to its plane. If, indeed, they had $\mathbf{k} \parallel \mathbf{B}$, the polarization phenomenon that they discovered should be regarded as the acoustic analog of the Faraday effect.

The first detailed report of this effect was published by Matthews and LeCraw in 1962 [9]. They also used YIG, but in a cylindrical form, 0.51 cm in diameter and 0.86 cm long. The axis of the specimen was parallel to the external magnetic field and to the [100] crystal axis, which is the magnetically hard direction for YIG. The experiment was performed at room temperature using a pulse-echo technique with a frequency of 528 MHz and a pulse duration of $2\,\mu\mathrm{s}$. An $AC$-cut quartz disk, bonded to the specimen faces with a silicone compound, served as a resonant transducer operating at very high overtones for generating and detecting ultrasonic vibrations with a well-defined direction, [100] in this case.

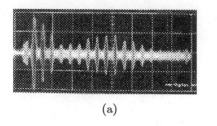 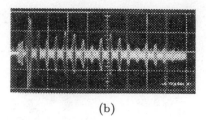

(a) (b)

FIGURE 5.1. Oscillograms of ultrasonic echo pulses detected by a piezoelectric quartz transducer at 528 Mhz in YIG with [100] $\|$ **H** $\|$ **k**. (a) $H = 2\,\text{kOe}$, (b) $H = 1.5\,\text{kOe}$. (After Fig. 1 in [9].)

Since the rotation of the polarization for the Faraday effect has odd symmetry with respect to inversion of the propagation direction, reflection of the ultrasonic pulse from the free end of the specimen and propagation backward to the transducer contributes to the total rotation the same amount as it does during forward propagation from the specimen–quartz interface. Thus, if one passage of a pulse through the specimen gives a polarization rotation of $\theta_1$, a round trip gives $2\theta_1$, so the rotation for the $n$th reflection (or "echo pulse") will be $\phi_n = 2n\theta_1$. This rotation was detected as a cosine variation of the amplitudes of successive echos displayed as a pulse train (see Fig. 5.1). The first node in signal amplitude occurs when the polarization plane has rotated $\pi/2$ radians. A node occurs for each additional rotation of $\pi$ radians.

The cosine variation is superimposed on the usual exponential decay caused by loss mechanisms. In YIG the loss is primarily caused by the magnon–phonon interaction. In the case of weak coupling this loss is manifested as ellipticity of the ultrasound, which is the reason that the minima do not go to zero in the oscillograms presented. Figure 5.2 shows the rotation values obtained by observing the applied fields at which particular echos are smallest.

The authors used these experimental results to calculate the magnetoelastic constant $b_2$. For this purpose, they took the dispersion equation obtained in [1], neglected higher-order coupling terms, and derived

$$\frac{k_+^2 - k_-^2}{k_0} = \frac{2\gamma\omega b_2^2/\rho M_0 s^2}{(\omega - \gamma H_i)(\omega + \gamma H_i) - [(2A/M_0)\,k_0^2]^2}. \tag{5.1}$$

Since $\phi = (k_- - k_+)l/2$ for weak coupling, they found

$$\frac{\phi}{l} = -\frac{\omega^2 b_2^2}{2\rho s^3 \gamma M_0}\left[\left(\frac{\omega}{\gamma}\right)^2 - (H - N M_0)^2\right]^{-1} \tag{5.2}$$

for $(2A/M_0)k_0^2 \ll \omega$, where $l$ is the distance traveled by the ultrasonic pulse. By fitting the data of Fig. 5.2(a) to equation (5.2) as shown in

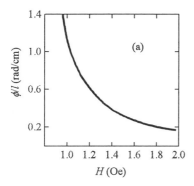
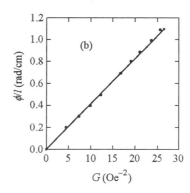

FIGURE 5.2. (a) Rotation of the polarization per unit length versus applied magnetic field in YIG. (b) Fit of experimental data to equation (5.2), where $G \equiv -10^7 \left[ \left( \frac{\omega}{\gamma} \right)^2 - (H - NM_0)^2 \right]^{-1}$. The slope of the straight line can be used to determine $b_2$. (After Figs. 2 and 3 in [9])

Fig. 5.2(b), the authors found the demagnetization field to be 360 Oe (in good agreement with that calculated for an ellipsoid of revolution with the same major and minor axes) and obtained $b_2 = 7.4 \times 10^6$ erg/cm$^3$ for the magnetoelastic constant.

*Important comments.* Equation (5.1) was derived from dispersion equations like (4.47)–(4.48) that describe the interactions of nondissipative subsystems. In general, there are three shear modes for resonance conditions with no energy loss: two coupled magnon–phonon modes having one circular polarization, and one phonon mode having the other circular polarization. Therefore equation (5.2), which takes only two modes into account, cannot be applied for conditions close to resonance. The coupled modes only become magnon-like and phonon-like outside a particular range of $H$ values. At the same time, a piezoelectric transducer is not an effective generator of a magnon-like mode far from resonance. Thus only two modes, a weakly coupled phonon-like mode with resonant polarization and a practically pure phonon mode with the opposite polarization, need to be considered. It appears that the authors had just this situation in their experiment. In the field range close to resonance there can be significant ellipticity of the ultrasound, causing uncertainty in the positions of the pulse amplitude minima, so the investigators were able to investigate the effects at higher fields better than near resonance.

In a later paper, Lüthi [10] presented the results of his ultrasonic studies of YIG. He used different mutual directions of polarization, wave vector, and applied magnetic field, and lower frequencies (50 to 150 MHz). His specimen was also in the form of a cylinder, 8 mm long and 3.5 mm in diameter. He found $b_2 = 19 \times 10^6$ erg/cm$^3$ by applying equation (5.2) to

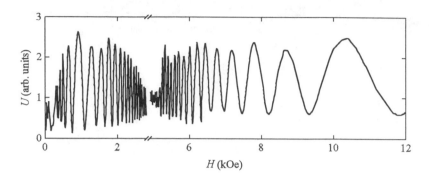

FIGURE 5.3. Signal amplitude in YIG at 9.375 GHz and 4.2 K. (After Fig. 2 in [13].)

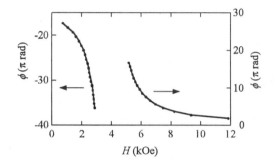

FIGURE 5.4. Rotation of the polarization in YIG at 9.375 GHz and 4.2 K. (After Fig. 3 in [13].)

Faraday effect data. By introducing spin-wave damping into the dispersion relation he obtained the spin-wave relaxation time $\tau_s \sim 10^{-9}$ s from the linewidth of the resonance.

Another investigation of the Faraday effect in YIG was carried out by Guermeur et al. [13]. They studied a 3 mm diameter, 9.5 mm long cylindrical rod with the [001] direction along the rod's axis. Apparently these dimensions provided a more uniform internal field in their sample compared with those used previously. Moreover, the ultrasonic frequency was much higher (9.375 GHz) and the temperature was 4.2 K. These improvements allowed the measurement of a very large rotation of the polarization (see Figs. 5.3 and 5.4). With the use of the method for determining $b_2$ introduced in [9], $13.5 \times 10^6$ erg/cm$^3$ was calculated for its magnitude, compared with the value of $14.4 \times 10^6$ erg/cm$^3$ that results from extrapolating the data of [9] to helium temperature.

Pavlenko et al. [14] also carried out an experiment in YIG that seems to have been the most accurate at that time. The advantage they had with respect to the earlier workers was the use of a YIG specimen having a

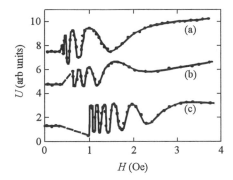

FIGURE 5.5. Amplitude of elastic pulses versus applied magnetic field for **H** ∥ [100] in a spherical YIG single crystal. Curves (a), (b), and (c) correspond to the second, fourth, and sixteenth pulses, respectively. The region of ultra rapid oscillations of the amplitude is indicated by dashed lines for Curves (b) and (c). Curves (a) and (b) have been shifted up by 6 and 4 units, respectively. (After Fig. 1 in [14].)

spherical shape. It was 8 mm in diameter and had small flat regions on opposite sides for placing $AC$-cut quartz transducers. The flat areas were rather small, so the distance between them was 7.5 mm. Linearly polarized waves were generated at a frequency of 820 MHz and propagated in the [100] direction. The data (some representative curves are shown in Fig. 5.5) were processed as in Ref. [9]. The resulting value for $b_2$ was $7.2 \times 10^6$ erg/cm$^3$.

Lemanov et al. [139] investigated spherical specimens of YIG with **k** and **H** along the [111] direction. Figure 5.6(a) shows the acoustic Faraday effect for linearly polarized waves. In addition, to prove that only one of the circularly polarized modes interacts resonantly with the spin waves, the authors placed a quarter-wave plate of yttrium aluminum garnet between the generating transducer and the specimen. The plate transformed the polarization of the wave from linear into circular. Figure 5.6(b) shows resonant interaction between the spin waves and the circularly polarized acoustic wave when the field is in one direction (triangles), but no interaction when the field is reversed (circles). In processing the data obtained for the [111] direction, it turned out to be possible to use equation (5.2), but with the substitution of the combination of two magnetoelastic constants, namely $b_{ef} \equiv (2b_1 + b_2)/3$, for $b_2$. They found $4.2 \times 10^6$ erg/cm$^3$ for the value of $b_{ef}$.

Tarasov et al. [140] also investigated ultrasonic propagation in YIG along the [001] direction; this was the first investigation of the Faraday effect where both ellipticity and rotation of the polarization were measured with the use of the precise phase-amplitude method described in Chapter 2. The experiments were performed with a spherical specimen at $T = 293.4$ K,

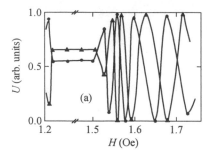 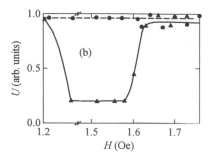

FIGURE 5.6. Amplitude of elastic waves propagating in YIG along [100] versus applied magnetic field: (a) Linearly polarized waves at 1340 MHz; (b) circularly polarized waves at 1500 MHz. Circles and triangles represent opposite field directions with respect to **k**. (After Fig. 3 in [139].)

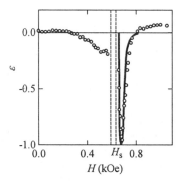 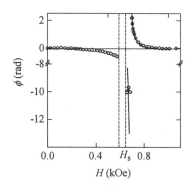

FIGURE 5.7. Magnetic field dependences of (a) the ellipticity and (b) the rotation of the polarization of ultrasound in a YIG single crystal with $\mathbf{k} \parallel \mathbf{H} \parallel [001]$ for $f = 58$ MHz. The vertical dashed lines indicate where the signal was too weak for measurements near the saturation field $H_s$. (After Fig. 3 of [140].)

using frequencies of 58, 177, 230, and 392 MHz. Taking advantage of their ability to obtain quantitative data on the widths of the resonant anomalies in ellipticity and rotation of the polarization (see Fig. 5.7), they derived information about energy dissipation in the magnetic subsystem for the single-domain state of the specimen. In addition, they discovered that in the inhomogeneous magnetic state below the field of magnetic saturation, $H_s$, there is a second magnon–phonon resonance for a field value $H_2$.

Using the resonant field data at for different frequencies they deduced the dependence of the spin wave frequency versus magnetic field in the vicinity of $H_s$ as shown in Fig. 5.8. The solid portions of Curves 1 and 2 indicate field ranges where the experimental results are consistent with the calculated field dependence of $\omega_s$, while the dashed portion is an extrapolation beyond the range of the measurements.

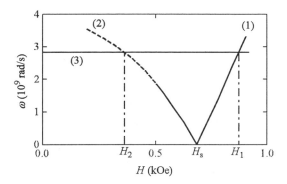

FIGURE 5.8. Magnetic field dependences of the spin wave $\omega_s$ (Curves 1 and 2) and the elastic wave $\omega$ (Curve 3) frequencies in a YIG single crystal with $\mathbf{k} \parallel \mathbf{H} \parallel [001]$. (After Fig. 1 of [140].)

## 5.2   The Cotton–Mouton Effect

The first paper on the acoustic analog of the Cotton–Mouton effect topic was published by Lüthi in 1963 [10]. We have already discussed the part of this paper dealing with the Faraday effect in Sec. 5.1. Recall that he used a cylindrical single crystal of YIG (3.5 mm in diameter and 8 mm long, probably [100]) at 50 to 150 MHz and at 300 K.

For this case the dispersion equation can be written as

$$\left(\omega^2 - s^2k^2\right)\left[\left(\omega^2 - s^2k^2\right)\left(\omega^2 - \omega_m{}^2\right) - \frac{\gamma^2 b_2^2}{\rho M_0}k^2 H_i\right] = 0, \tag{5.3}$$

where

$$\omega_m{}^2 = \gamma^2 H_i\left(H_i + \frac{3K_1}{M_0} + 4\pi M_0\right), \tag{5.4}$$

the internal field is assumed to be $H_i = H - N M_0 - (2K_1/M_0)$, and $K_1$ is the cubic first-order anisotropy constant.

Only the shear wave with polarization parallel to the magnetic field is coupled to the spin wave, as confirmed by the experimental data shown in Fig. 5.9. Using a method similar to the one developed while investigating the Faraday effect [9], Lüthi found the magnetoelastic constant $b_2 = 2.4 \times 10^7$ erg/cm$^3$ from Curve (c). At the same time, by using the linewidth and the height of the resonance Curve (a), the relaxation time was calculated to be $\tau_s = 3 \times 10^{-9}$ s. The was the first suggestion of the possibility of studying the Cotton–Mouton effect with a method developed for the Faraday effect.

A more complete description of this method was given in the next paper by Lüthi [141] on gadolinium-iron garnet (GdIG) at its compensation point at room temperature. The long axis of the specimen was along [100], the

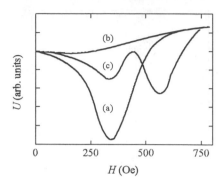

FIGURE 5.9. Acoustic amplitude versus magnetic field applied perpendicular to wave propagation at 70 MHz in YIG. (a) Wave polarization parallel to **H**, (b) perpendicular to **H**, and (c) between these two directions, corresponding to the Cotton–Mouton effect. (After Fig. 1 in [10].)

waves were polarized along [001], and the magnetic field was along [011]. In this geometry the component of the shear wave that is parallel to the magnetization couples to the spin wave whereas the other, which is perpendicular to it, does not. Therefore, one of the wave vectors has a magnetic field dependence while the other does not. This leads to a phase difference per unit length

$$\frac{\Delta\varphi}{l} = \frac{\gamma^2 H_i \omega b_2^2}{\rho s^3 M_0 \left(\omega_m^2 - \omega^2\right)}. \tag{5.5}$$

The phase shift leads to elliptical polarization of the vibrations at the receiving transducer. In turn, elliptical vibrations cause minima in the signal amplitude when the ellipticity is equal to $\pm 1$ $[\Delta\varphi = \pi(n + \frac{1}{2})$, circular polarization, where $n = 0, \pm 1, \pm 2, \ldots]$ and maxima for $\varepsilon = 0$ $(\Delta\varphi = \pi n$, linear polarization). By using subsequent maxima and minima, one can plot $\Delta\varphi/l$ versus $H$ for all of the echoes. Experimental values for $K_1$, $b_2$, and $\gamma$ can be obtained by fitting these values to equation (5.5) if the other parameters are known. Figure 5.10 shows the results of such a fitting.

In a subsequent paper Lüthi [11] used the Cotton–Mouton effect in magnetite ($Fe_3O_4$) and nickel to measure the constant $b_2$. Because of the large ultrasonic attenuation in nickel, he could only obtain an upper limit for $|b_2|$ of $2 \times 10^8$ erg/cm³ at 77 K. He applied the generalized form of equation (5.5) to magnetite for the case when both the first- and the second-order anisotropy constants are important,

$$\frac{\Delta\varphi}{l} = \frac{\omega b_2^2}{\rho s^3 M_0} \frac{H_i}{H_i \left(H_i + \frac{3K_1}{M_0} + \frac{K_2}{2M_0} + 4\pi M_0\right) - \left(\frac{\omega}{\gamma}\right)^2}, \tag{5.6}$$

to fit the data, and obtained $|b_2| = 1.6 \times 10^8$ erg/cm³ at room temperature.

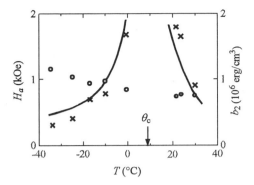

FIGURE 5.10. Magnetic crystallographic anisotropy field $H_a = -2K_1/M_0$ as a function of temperature $T$ in GdIG (solid curve: calculated; crosses: measured.) The circles are the values of $b_2$ found by fitting the experimental data to equation (5.5). (After Fig. 3 in [141].)

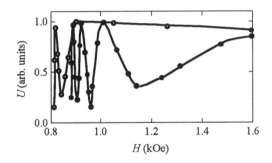

FIGURE 5.11. Pulse amplitude versus magnetic field for waves propagating along [110] in YIG. Empty circles are for $\mathbf{H} \parallel [\bar{1}10]$, full circles for $\mathbf{H} \parallel [001]$. (After Fig. 6 in [139].)

In addition to the Faraday effect measurements in spherical specimens reported in the preceding section, Pavlenko et al. also observed the Cotton–Mouton effect in the same samples [14]. It gave a value of $6.8 \times 10^6$ erg/cm$^3$ for the magnetoelastic constant $b_2$, which is slightly smaller than the value from the Faraday effect.

Taking advantage of the fact that YIG is practically isotropic in its elastic properties, Lemanov et al. [139] investigated it using a transverse wave propagating along [110], parallel to the magnetic field. The polarization direction was 45° from [001], so both [001] and [$\bar{1}$10] components were excited. Figure 5.11 shows the amplitude dependence on magnetic field strength for two mutually perpendicular directions of $\mathbf{H}$. For these two field directions the authors derived the following formula for phase shift

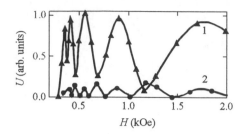

FIGURE 5.12. Magnetic field dependence of the pulse amplitude of transverse waves propagating along the hexagonal axis in the ferromagnet RbNiF$_3$. The magnetic field is in the $X$-$Y$ plane, perpendicular to $\mathbf{k}$, and $T = 77\,\mathrm{K}$. Curve 1 is for the first pulse and Curve 2 for the second. (After Fig. 1 in [142].)

between the two lattice displacement components:

$$\frac{\Delta\varphi}{l} = \frac{\pi\nu b_{ef}^2 H}{\rho s^3 M_0}\left[H\left(H + 4\pi M_0\right) - \left(\frac{\nu}{\gamma}\right)^2\right]^{-1}. \tag{5.7}$$

They used this equation to analyze their data and found $b_1 = 3.8 \times 10^6\,\mathrm{erg/cm^3}$ and $b_2 = 6.7 \times 10^6\,\mathrm{erg/cm^3}$, where $b_1$ stands for $b_{ef}$ with $\mathbf{H} \parallel [\bar{1}10]$ and $b_2$ for $b_{ef}$ with $\mathbf{H} \parallel [001]$.

In contrast to the works discussed above which dealt with cubic crystals, Grishmanovskii et al. [142] investigated the Cotton–Mouton effect in the hexagonal ferromagnet RbNiF$_3$ ($D_{6n}$ point group). The samples were in the form of parallelepipeds measuring $8 \times 8 \times 3\,\mathrm{mm^3}$. They used a frequency of about $800\,\mathrm{MHz}$ and temperatures from 77 to $140\,\mathrm{K}$. The waves propagated along the $Z$ axis, and the external magnetic field was in the $X$-$Y$ plane and made an angle of $\pi/4$ with the initial polarization of the waves. Some of the data obtained are shown in Figs. 5.12 and 5.13.

To describe the phenomena investigated, the authors used the following forms for the equations for the magnetoelastic and magnetocrystallographic anisotropy energies:

$$W_{me} = b_{ijkl}\varepsilon_{ij}\alpha_k\alpha_l, \qquad W_a = K_1\alpha_z^2 + K_2\alpha_z^4. \tag{5.8}$$

For such a crystal there are five independent components of elastic moduli, $c_{ijkl}$, and four magnetoelastic constants, which may be written as $b_{11} - b_{12}$, $b_{13} - b_{12}$, $b_{33} - b_{31}$, and $b_{44}$.

For wave propagation along the hexagonal axis the authors used the following solutions of the dispersion equation:

$$k_1^2 = \omega^2\left[s_z^2 - \frac{\frac{4\gamma^2}{\rho M_0}b_{44}^2 H}{\gamma^2 H\left(H + 4\pi M_0 + \frac{2K_1}{M_0}\right) - \omega^2}\right]^{-1}, \qquad k_2^2 = \frac{\omega^2}{s_z^2}, \tag{5.9}$$

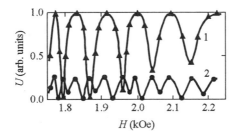

FIGURE 5.13. Magnetic field dependence of the pulse amplitude of transverse waves propagating along the $X$ axis in RbNiF$_3$. The magnetic field is in the $X$-$Y$ plane, perpendicular to $\mathbf{k}$, and $T = 77\,\mathrm{K}$. Curve 1 is for the first pulse and Curve 2 for the second. (After Fig. 2 in [142].)

where $s_z = (c_{2323}/\rho)^{\frac{1}{2}}$ is the shear wave velocity, which was measured as $2.7 \times 10^5$ cm/s.

For propagation along the $X$ axis,

$$k_1^2 = \omega^2 \left[ s_{x1}^2 - \frac{\frac{4\gamma^2}{\rho M_0}(b_{11}-b_{12})^2 \left(H+\frac{2K_1}{M_0}\right)}{\gamma^2 \left(H+\frac{2K_1}{M_0}\right)(H+4\pi M_0)-\omega^2} \right]^{-1} , \quad k_2^2 = \frac{\omega^2}{s_{x2}^2}, \quad (5.10)$$

where $s_{x1} = [(c_{1111}-c_{1122})/2\rho]^{\frac{1}{2}}$, $s_{x2} = (c_{2323}/\rho)^{\frac{1}{2}}$. Their measurements yielded $s_{x1} = s_{x2} = 2.7 \times 10^5$ cm/s. Here and in equations (5.9) the indices 1 and 2 denote waves polarized along and perpendicular to the field, respectively.

Using the solutions of (5.9) and (5.10), the equations for the phase difference between waves of different polarization can be derived. For the first case,

$$\frac{\varphi}{l} = \frac{4\pi\nu}{\rho s_z^3 M_0} b_{44}^2 \left[ H\left(H+4\pi M_0 + \frac{2K_1}{M_0}\right) - \left(\frac{\nu}{\gamma}\right)^2 \right]^{-1} . \quad (5.11)$$

For the second,

$$\frac{\varphi}{l} = \frac{4\pi\nu}{\rho s_x^3 M_0} (b_{11}-b_{12})^2 \frac{\left(H+\frac{2K_1}{M_0}\right)}{\left[\left(H+\frac{2K_1}{M_0}\right)(H+4\pi M_0)-\left(\frac{\nu}{\gamma}\right)^2\right]} . \quad (5.12)$$

Using their data and values for some constants taken from other publications (see references in [142]), the authors found $b_{44} = 2.7 \times 10^7$ erg/cm$^3$ at 77 K. They also measured the temperature dependence of this parameter, as shown in Fig. 5.14.

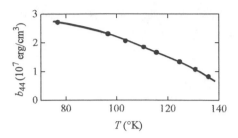

FIGURE 5.14. Temperature dependence of the magnetoelastic constant $b_{44}$ in RbNiF$_3$. (After Fig. 4 in [142].)

Burkhanov et al. studied the Cotton–Mouton effect in polycrystalline NiFe$_2$O$_4$ [81]. The experiment was performed at 78 K using a frequency of 51 MHz in fields corresponding to a magnetically saturated specimen as well as to a polydomain condition. This paper was the first where the effect was investigated with the precise technique described in Sec. 2.3. Moreover, to give the reader an idea of how ellipticity and rotation of the polarization depend on the differences between the wave numbers $\Delta k$ and the absorption $\Delta \Gamma$ of linearly polarized normal modes [see equations (2.7) and (4.104)], the authors presented $\varepsilon$ and $\phi$ as three-dimensional surfaces, assuming that the effect is due to linear birefringence and dichroism of ultrasound. The following equations were derived for the polarization parameters:

$$\phi = \varphi - \Psi, \qquad \tan 2\varphi = \frac{2R}{1 - R^2} \cos\left(\Delta kl\right),$$

$$\varepsilon = \frac{1 - |q|}{1 + |q|}, \qquad q^2 = \frac{\left(1 + R^2\right) + 2R \sin\left(\Delta kl\right)}{\left(1 + R^2\right) - 2R \sin\left(\Delta kl\right)}, \qquad (5.13)$$

where $R = (\tan \Psi) \exp\left(\Delta \Gamma\right)$, $\Delta k = \mathrm{Re}\left(k_1 - k_2\right)$, $\Delta \Gamma = \left(\Gamma_1 - \Gamma_2\right)$, $k_1$ and $k_2$ are components of the wave vectors $\mathbf{k}_i = (0, 0, k_i)$ relating to normal modes with polarization parallel and perpendicular to the magnetic field $\mathbf{H} = (H, 0, 0)$, $l$ is the distance traveled by the wave, $\Psi$ is the angle between the magnetic field and the initial direction of the polarization plane, and $\varphi$ is the angle between the magnetic field and the major axis of the polarization ellipse for $H \neq 0$.

Figure 5.15 presents only portions of the complete surfaces. They can easily be continued: $\varepsilon$ is an odd function of $R$, vanishing as $R \to 0$, and an odd periodic function of $\Delta kl$, while $\varphi$ is an odd function of $R$ and a periodic even one of $\Delta kl$. These images have wider application than just illustrating the effects in a particular material: they display the complete variety of the magnitudes that can be obtained in an experiment. Those observed in NiFe$_2$O$_4$ are shown as curves on these surfaces. Both of the curves have the same projection on the plane $(\Delta kl, R)$ (see Fig. 5.16).

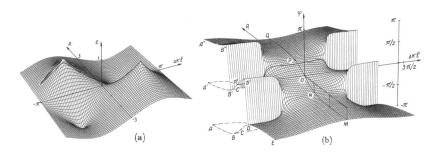

FIGURE 5.15. (a) Ellipticity and (b) rotation of the polarization due to linear birefringence and dichroism of ultrasound. The curves on the surfaces relate to $\varepsilon$ and $\varphi = \phi + \Psi$ obtained in $NiFe_2O_4$ in the magnetic field interval $-0.3\,\text{kOe} \le H \le 1\,\text{kOe}$. (After Figs. 3 and 4 in [81].)

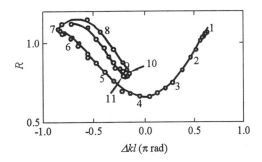

FIGURE 5.16. Effect of the magnetic field on the parameters characterizing dichroism ($R$) and birefringence ($\Delta kl$) in $NiFe_2O_4$. The numbers refer to the following values of $H$: (1) $\pm 1.0$, (2) $\pm 0.65$, (3) $\pm 0.5$, (4) $\pm 0.4$, (5) $\pm 0.3$, (6) $\pm 0.2$, (7) $\pm 0.15$, (8) $+0.1$, (9) $+0.08$, (10) $+0.06$, and (11) $0\,\text{kOe}$. (After Fig. 5 in [81].)

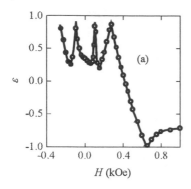 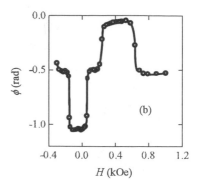

FIGURE 5.17. (a) Ellipticity and (b) rotation of the polarization versus magnetic field in $NiFe_2O_4$ for $f = 51\,MHz$, $T = 78\,K$, $l = 9.5\,mm$. (After Figs. 1 and 2 in [81].)

The magnetic field dependences of the ellipticity and the rotation of the polarization are presented in Fig. 5.17. One can see that appreciable changes in the value of $\phi$ take place only near certain values of $H$, being almost constant over wide intervals. This behavior, as well as the fact that the ellipticity is equal to $\pm 1$ in its extrema, are typical for an effect that is almost entirely due to linear birefringence. This conclusion is supported by the curve in Fig. 5.16: It remains near $R = 1$ while $\Delta kl$ changes from approximately $-3/4\pi$ to more than $+\pi/2$. The large variations in phase velocity in this material are due to two mechanisms for magnetoelastic interaction which were discussed in [143]; one is independent of the magnetization homogeneity and the other is related solely to the multiple-domain state.

## 5.3   The Kerr Effect

A change in the polarization of transverse ultrasonic waves upon reflection from a magnetic-non magnetic media interface was found by Burkhanov et al. and reported in brief in [144]. A more complete description of the effect and its interpretation are given in [145]. The authors used a prism of fused quartz as the isotropic non magnetic material and a crystal of YIG as the magnetic specimen. Figure 5.18 shows the arrangement of the prism and the specimen. According to the classification given in Sec. 4.6, this was an investigation of the polar Kerr effect with a $p$-type incident wave.

The YIG specimen (used in the earlier experiments on the Faraday effect [140]) had the form of an 8.4 mm diameter sphere with small, flat regions for attaching transducers. The line width of the ferromagnetic resonance in this sample in its initial spherical form was 0.5 Oe. The prism and the

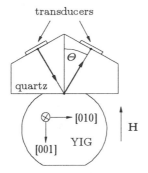

FIGURE 5.18. Arrangement of a quartz prism and a YIG single crystal for observing the polar Kerr effect. Arrows indicate the generating and receiving piezoelectric transducers. (After Fig. 1 in [145].)

sample were bonded together with an epoxy resin. The thickness of the bond was less than a wavelength of light. The external magnetic field was perpendicular to the interface and parallel to the [001] crystallographic axis. The plane of polarization of the incident wave was in the (100) plane of the YIG specimen, and the angle of incidence, $\theta$, was 22.5°.

The ellipticity and rotation of the polarization were measured with the amplitude technique described in Secs. 2.3.2 and 2.4. Figure 5.19 shows the results of the experiment after processing the data measured for three values of the angle $\psi$.

In order to compare the experimental results with theory, calculations based on the scheme presented in Sec. 4.6.2 were carried out, with results as shown in Fig. 5.20. Remember that this scheme includes the assumption that the sample is in a magnetically homogeneous state, thus the results are valid only above the field of magnetic saturation, $H_S = 667$ Oe. It follows that only the high-field wing of the curves can be observed at the frequency used in the experiment with the relaxation parameter $\alpha_0 > 0.1$.

Comparing the experimental and theoretical curves one notes that the experimental ones consist of a set of resonances with different widths superimposed on a slowly varying background, in contrast with the typical resonant curves of $\varepsilon(H)$ and $\phi(H)$ obtained in the Faraday effect [140]. The principal difference between these two effects is the following. When a wave propagates through a bulk specimen, its characteristic properties (phase velocity, absorption, polarization, etc.) are determined by the bulk parameters of the medium, but the reflection coefficients depend upon properties of the media in a region near the interface with a thickness of the order of the wavelength. Only in the most primitive and commonly discussed case of isotropic, homogeneous media are the reflection and transmission coefficients expressible in terms of bulk parameters. Taking this circumstance

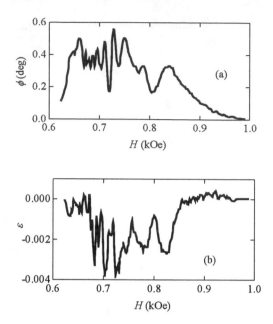

FIGURE 5.19. (a) Rotation of the polarization and (b) ellipticity of 53.5 MHz transverse waves reflected from the interface between a YIG crystal and fused quartz at $T = 295$ K. (After Fig. 3 in [145]).

into account, the authors suggested the following interpretation of what was observed.

In a real magnetic crystal there can be elastic stresses that result in an additional field caused by magnetic anisotropy. The stresses inside a high-quality crystal of YIG are very small, but they can be significant on the flat areas since there was no thermal processing after they were made on the initially spherical sample. Another source of elastic stresses is magnetostriction. The bonding of the sample and the prism was done with $H = 0$. All of the experiments were carried out after the bond hardened. When a magnetic field is applied perpendicular to the interface, the dimensions of the flat region tend to increase. While firmly attached to the quartz, whose dimensions are independent of $H$, the sample experiences forces of compression which depend on the external magnetic field, according to the magnetostriction data obtained by Callen et al. [146]. Owing to the finite dimensions of the contact area, the stresses are nonuniform and should at least cause broadening of the resonances.

In the area of contact with the prism the sample differs the most from a spherical shape. Here the demagnetizing field is also nonuniform, which is another cause of broadening of the resonances. Actually, the parameter $\alpha_0$ introduced as the relaxation parameter becomes the characteristic

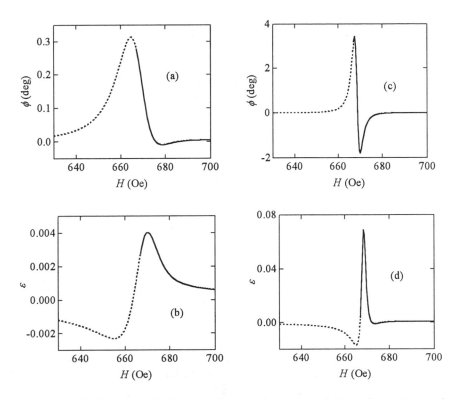

FIGURE 5.20. Magnetic field dependences of rotation of the polarization and ellipticity of ultrasound reflected from the interface between a YIG crystal and fused quartz calculated for 53.5 MHz and an angle of incidence of $\theta = 22.5°$. The relaxation parameter $\alpha_0$ has been taken to be 0.5 in (a) and (b), and 0.1 in (c) and (d). The orientations of the magnetic field and crystallographic axes are as in Fig. 5.18. The dotted portions of the curves correspond to the non-homogeneous magnetic state of the YIG sample, the solid portions to the magnetically saturated state. (After Fig. 5 in [145].)

property of inhomogeneous resonance and it is not associated directly with absorption of the spin waves.

And, finally, the resonance was observed in the vicinity of $H_S$. The stresses mentioned above prevent the bonding surface of the sample from transforming into a magnetically saturated state. It can be assumed that the region where the ultrasonic wave reflects consists of a number of domains with magnetization close to, but deviating slightly from, the equilibrium value. In this case the resonance condition may vary in different domains, being distributed about the magnetic field determined for a uniform magnetic state.

Since the reflection of ultrasound from different domains is characterized by different local values of $\varepsilon$ and $\phi$, the values determined in the experiment represent averages over the region where the wave reflects. These values will be designated by $\bar{\varepsilon}$ and $\bar{\phi}$. In turn, they are calculated using the averages $\overline{N}_{11}$, $\overline{N}_{21}$, and $\overline{N}_{31}$. In general, the functions used for the calculations will differ from those given in Sec. 2.3.2. However, if $\bar{\varepsilon}$ and $\bar{\phi}$ are small, the functions for their calculation can be expanded in a series and then treated in the linear approximation for $N_{11}$, $N_{21}$, and $N_{31}$. In such a representation, constant parameters of the polarization of ultrasound and averaged ones will have identical functional dependences on $N_{11}$, $N_{21}$, and $N_{31}$, or $\overline{N}_{11}$, $\overline{N}_{21}$, and $\overline{N}_{31}$. Since the observed values of $\varepsilon$ and $\phi$ were indeed small, the quantitative data shown in Fig. 5.19 are valid even for an inhomogeneous magnetic state at the interface. The contribution of a certain domain to $\overline{N}_{11}$, $\overline{N}_{21}$, and $\overline{N}_{31}$ is proportional to the square of its relative area on the reflecting surface. As a result, the modifications of a resonance caused by the contributions of the domains to $\bar{\varepsilon}$ and $\bar{\phi}$, having the same width in $H$, are reduced with respect to the values that would be observed for a homogeneous magnetic state at the interface.

Therefore, the quite narrow resonances (with width approximately equal to $10-15$ Oe in spite of the small values of $\varepsilon$ and $\phi$) correspond to $\alpha_0 \approx 0.2$ and obviously are due to the domains from the central part of the reflecting area. The background of the curves is caused by the part of the area away from the center where the magnetic field is the most inhomogeneous. The relaxation parameter characterizing this part of the interface is much larger and its contribution to the resonant curves is represented only in the high-field wings.

In comparing the application of the Kerr and Faraday effects to ferromagnets it should be noted that in the first phenomenon the surface stresses and the domains caused by these stresses are of great importance. They lead to the main differences in the data obtained from the two effects, even when the same specimen is investigated. The bulk phenomenon can be described quite well with a theory that assumes a homogeneous magnetic state, while the reflection of ultrasound requires a more accurate description of the magnetic state at the interface. Consequently, information obtained from the Kerr effect will relate entirely to this interface.

# 6
# Experiments in Nonmagnetic Metals

## 6.1 The Faraday Effect: Doppler-Shifted Cyclotron Resonance

The rotation of the polarization and the ellipticity of ultrasound for DSCR are formally described with the aid of the wave vectors obtained in the weak-coupling approximation and given by equation (4.92). In this equation the electroacoustic coefficients should be written as functions of $\mathbf{k}_0$ instead of $\mathbf{k}$ since the equation results from the first iteration in the solution of the dispersion equation (4.77). The coefficients $\alpha^\pm$, $\sigma^\pm$, and $\beta^\pm$ are defined by (4.80)–(4.81), and $\mathbf{k}_0$ has only an $H$ component as we are discussing the Faraday effect with both $\mathbf{k}$ and $\mathbf{H}$ parallel to the $z$ axis. In this chapter we will omit the subscript $e$ which we used earlier to indicate that $\Gamma$, $s$, and $\mathbf{k}$ refer to elastic-like modes.

Each of the electroacoustic coefficients, according to (4.72), contains the resonant denominator $\nu - i\left(\mathbf{k}\cdot\bar{\mathbf{v}} - \omega - n\Omega\right)$ in which $\Omega$ and $\mathbf{k}\cdot\bar{\mathbf{v}}$ play the most important roles [recall that $\Omega$ is defined as the cyclotron frequency by equation (4.58), and $\mathbf{k}\cdot\bar{\mathbf{v}}$ is the Doppler shift of the frequency caused by the motion of an electron along the magnetic field with average velocity $\bar{\mathbf{v}}$]. The integer $n$ before the $\Omega$ appears if the cyclotron orbit is not circular, in which case the motion can be represented by a sum of cyclic harmonics with a fundamental resonance at $n = \pm 1$. We will only discuss the cases when resonance manifests itself rather clearly, i.e., when

$$(\mathbf{k} \cdot \bar{\mathbf{v}})\tau \sim k\ell \gg 1 \qquad \text{and} \qquad \Omega\tau \geq 1. \qquad (6.1)$$

The first condition holds when the electron mean free path is much longer than the ultrasonic wavelength, and the second when the resonant electrons are able to complete an orbit before scattering.

When $\omega\tau \ll 1$, the value of $H$ at which a group of electrons with cyclotron mass $m_c$ and velocity $\bar{\mathbf{v}}$ cause a fundamental DSCR in a particular electroacoustic coefficient is given by

$$H = \left| \frac{\mathbf{k}_0 \cdot \bar{\mathbf{v}}\, m_c\, c}{e} \right| = \frac{k_0 c}{2\pi|e|} \left| \frac{\partial S(p_z)}{\partial p_z} \right| = \frac{k_0 c h}{4\pi^2 |e|} \left| \frac{\partial S(\kappa_z)}{\partial \kappa_z} \right|, \qquad (6.2)$$

where $\kappa_z$ is the $z$ component of the wave vector of the charge carriers.

As the electrons in a metal have a quasi-continuous but finite range of $m_c$ and $\mathbf{v}$ values, there is an upper limit to the value of $H$ for which DSCR can occur. It is called the Kjeldaas field (or Kjeldaas edge) [106] $H_K$, or the cyclotron absorption edge. In a free-electron metal, the FS is spherical so $\left|\frac{\partial S(\kappa_z)}{\partial \kappa_z}\right| \sim |\kappa_z|$ and $H_K$ is directly proportional to the Fermi wave vector $\kappa_F$. In more complicated metals the FS will have a more complicated form, making it difficult to relate $H_K$ to a single feature.

$H_K$ should not be regarded as the upper limit of all electronic attenuation. In some metals magnetoacoustic effects caused by conduction electrons can also be observed above $H_K$, such as giant quantum oscillations of ultrasonic attenuation [147, 148]. Those are related to the zeroth harmonic (the harmonic with $n = 0$) and are not the result of a resonance.

Equation (6.2) shows that parameters of the FS such as the derivative $\partial S/\partial p_z$ (or $\partial S/\partial \kappa_z$), which is proportional to $m_c|\bar{v}_z|$, can be measured while investigating DSCR in the most typical situation. However, if $\omega\tau \geq 1$, resonances caused by electrons with opposite velocities $\pm\bar{v}_z$ occur at slightly different fields $H_\pm$, permitting $m_c$ and $|\bar{v}_z|$ to be determined separately:

$$m_c = \frac{|e|}{2c\omega}\,(H_+ - H_-), \qquad \bar{v}_z = s_0 \frac{H_+ + H_-}{H_- + H_-}. \qquad (6.3)$$

Conduction electrons located in so-called *effective zones* on the FS play the most significant role in magnetoacoustic phenomena in metals. With regard to DSCR they are in the form of belts defined by $p_H = \text{const}$, or in our case, $p_z = \text{const}$, having a width $\Delta p_z/p_z \sim 1/k_0\ell$ as a rule. All of the observable irregularities in the magnetic field dependences of ultrasonic absorption, phase velocity, ellipticity, and rotation of the polarization are due to belts corresponding to either extrema of $\partial S/\partial p_z$ or limiting points of the FS. Thus, the parameters given by (6.3) can be determined for particular belts of effective electrons when performing experiments in a longitudinal magnetic field.

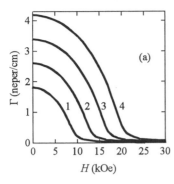 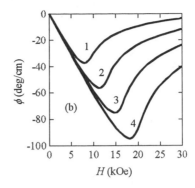

FIGURE 6.1. Calculated (a) attenuation and (b) rotation of the polarization of shear waves versus magnetic field in free-electron copper assuming $k_0\ell = 10$ at 50 MHz. The frequency $f = 50$ Mhz, 70 Mhz, 90 MHz, and 110 MHz for Curves $1 - 4$, respectively. (After Figs. 2 and 3 in [20].)

Rotation of the polarization of ultrasound was first observed by Morse and Gavenda [7] in 1959 under DSCR conditions in copper. At that time it was still unknown whether there was a cyclotron absorption edge in metals, thus a number of experiments were aimed at determining the magnetic field dependence of ultrasonic absorption and, in particular, the form of its drop at $H_K$. These authors remarked that the attenuation at high frequencies did not diminish with increasing $H$ as rapidly as the theory predicted. They conjectured that this was caused by rotation of the polarization away from the sensitive direction of the receiving transducer. The rotation should be greatest at the Kjeldaas field for a metal with a free-electron FS as shown in Fig. 6.1.

Later, Jones [16] performed experiments in an aluminum specimen having residual resistance ratio (RRR) $R_{300K}/R_{4.2K} = 10\,400$ with **H** and **k** along $C_4$ using transmitting and receiving transducers that were rotated with respect to one another by $\varphi = \pi/4$ in the plane perpendicular to **k**. He found differences in the graphs of signal loss versus $1/H$ caused by reversing the field direction as shown in Fig. 6.2. The different behavior for the two different field directions can be explained as being the result of odd parity for the Faraday effect under reversal of the applied magnetic field. If a field parallel to **k** rotates the polarization toward the sensitive direction of the receiving transducer, the reversed field will rotate the polarization away, resulting in a smaller signal. The differences displayed in Fig. 6.2 were the first clear evidence for the existence of magnetic field-induced rotation of the polarization in metals.

The first quantitative measurements of $\phi(H)$ in a metal were obtained in copper by Gavenda and Boyd [19] after developing the method described in [20] for precisely measuring the rotation of the plane of polarization. They used a receiving transducer with its polarization direction rotated approxi-

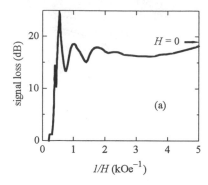 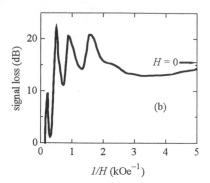

FIGURE 6.2. Magnetic field dependence of the signal loss for 50 MHz shear waves propagating along [100] in aluminum when the polarization of the receiving transducer is rotated by an angle of $\pi/4$ with respect to that of the generating one. (a) $\mathbf{k} \parallel \mathbf{H}$. (b) $\mathbf{k} \parallel -\mathbf{H}$. The crystal thickness is 0.818 cm and $T = 4.2$ K. (After Fig. 5 in [16].)

mately $\pi/4$ from that of the transmitting transducer and observed the variations in received signal for magnetic field both parallel and anti-parallel to $\mathbf{k}$ (as in [16]) for frequencies in the range 31 to 110 MHz. Figure 6.3 illustrates the different behaviors of the signal amplitude for the two cases ($\mathbf{k} \parallel \pm\mathbf{H}$) with $\mathbf{k} \parallel [001]$ in a copper crystal having electron relaxation time $\tau \approx 5 \times 10^{-11}$ s at 4.2 K.

These authors separated out the effects caused by polarization rotation from those caused by absorption through the use of a model based on the fact that the receiving transducer is sensitive only to the component of lattice vibration parallel to its polarization. Some results of this analysis of the data are shown in Fig. 6.4. These results are plotted against inverse field strength to make evident their periodic oscillatory character. In order to explain the great differences from the calculations shown in Fig. 6.1 for the free-electron model, they found it necessary to take into account the properties of the copper Fermi surface, using the empirical model proposed by Roaf [149]. The FS parameters that play major roles in attenuation and rotation phenomena are plotted in Fig. 6.5.

Although one can see an absorption edge in the plots of attenuation versus $f/H$, it is not of the simple type discussed by Kjeldaas; there is a resonant peak below the edge, as predicted by Kaner et al. [150], along with harmonic oscillations caused by the noncircularity of the orbits centered at $\kappa_z = 0.45 \times 10^8$ cm$^{-1}$ that contribute to the resonant peak. However, all of these features were explained in terms of the shape of the copper FS. Using a relatively crude model for calculating the attenuation, the authors obtained good agreement with the location of the fundamental resonance and the general shape of the measured attenuation, as seen in Fig. 6.6. The lower horizontal line in Fig. 6.5(c) represents the value of $H$

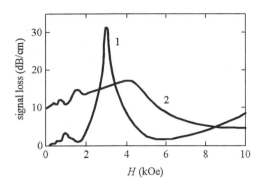

FIGURE 6.3. Signal loss in copper as a function of magnetic field when the polarization direction of the receiving transducer makes an angle $\varphi \approx \pi/4$ with that of the generating one. Curve 1 corresponds to $\mathbf{k} \parallel \mathbf{H}$ and Curve 2 to $\mathbf{k} \parallel -\mathbf{H}$. Frequency 92 MHz, $T = 4.2$ K, wave propagation along [001]. (After Fig. 6 in [20].)

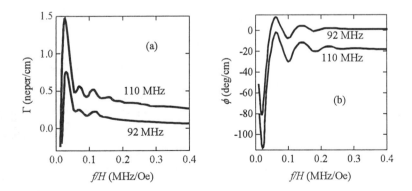

FIGURE 6.4. (a) Attenuation and (b) rotation of the polarization in copper with $[001] \parallel \mathbf{k} \parallel \mathbf{H}$ at 4.2 K. (After Figs. 9 and 11 in [20].)

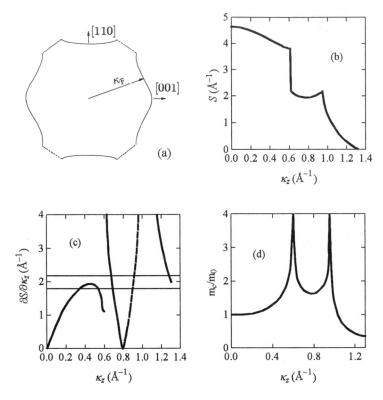

FIGURE 6.5. Properties of the copper Fermi surface based on Roaf's model [149]. (a) Central cross section in the (110) plane. The dashed lines represent the intersection of the FS with the Brillouin zone boundaries, and the Fermi wave vector in the [001] direction is $\kappa = 1.33 \times 10^8 \, \text{cm}^{-1}$. (b) Cross-sectional area versus $\kappa_z$ for $\kappa_z$ in the [001] direction. (c) $|\partial S/\partial \kappa_z|$ versus $\kappa_z$ in the [001] direction. The dashed line indicates where it is positive, the solid line where it is negative. (d) Cyclotron mass for $\mathbf{H}$ along [001] normalized with respect to its value at $\kappa_z = 0$. (After Figs. 14–17 in [20].)

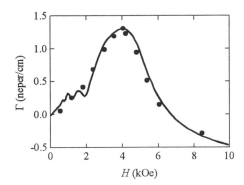

FIGURE 6.6. Comparison of the attenuation measured in copper (solid line) with that calculated for 110 MHz (filled circles). (After Fig. 18 in [20].)

for the fundamental resonance, which should correspond to a maximum in $|\partial S/\partial \kappa_z|$. The upper horizontal line in the same figure is obtained from the period of the harmonic oscillations in the attenuation and rotation data. It should agree with the value found from the fundamental resonance. The discrepancy is likely due to the fact that $\Omega \tau$ was not very large in these experiments.

In fitting the data, they found that the major contributions to the main peak in the attenuation come from electrons with $0.2 \times 10^8 \, \text{cm}^{-1} < \kappa_z < 0.58 \times 10^8 \, \text{cm}^{-1}$. For these electrons they obtained $m_c \overline{v_z} = 0.381 \pm 0.009 \times 10^{-19} \, \text{g cm/s}$ and $\partial S/\partial \kappa_z = -2.28 \pm 0.06 \times 10^8 \, \text{cm}^{-1}$.

Miller [23] made similar measurements in tin and aluminum at frequencies of 60 and 84 MHz. He used two adjacent receiving transducers having their polarization directions $\pm \pi/4$, respectively, from the polarization direction of the generating transducer. The generated linearly polarized wave is equivalent to two oppositely rotating circularly polarized waves. The signal from one of the receiving transducers was delayed by one-fourth period before being combined with the other so that the detected sum corresponded to one of the circularly polarized modes. Appreciable differences between the absorption curves for the two circularly polarized modes were observed in tin (RRR = 50 000), but only very small differences occurred in aluminum (RRR = 5 000) (see Fig. 6.7).

In addition, Miller measured the phase shift between the two circularly polarized components, $\Delta = (k^+ - k_-)l/2$, where $l$ is the ultrasonic path length, which corresponds to our definition of the angle of rotation of the polarization plane under weak coupling [see equation (4.103)]. Figure 6.8 shows the curve of $\Delta(H)$ obtained in tin.

Hui and Rayne [24] also studied DSCR in aluminum. They had a purer specimen (RRR = 14 800), used a wide range of frequencies (30 to 210 MHz), and investigated three orientations: field and wave-vector directions along

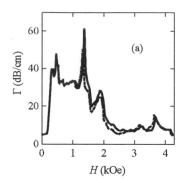 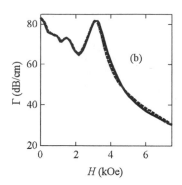

FIGURE 6.7. Attenuation of $(+)$ (dotted lines) and $(-)$ (solid lines) circularly polarized waves propagating along [001] (a) in tin and (b) in aluminum. $T = 1.2\,\mathrm{K}$ for both. (After Figs. 4 and 7 in [23].)

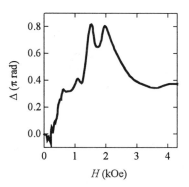

FIGURE 6.8. Phase shift $\Delta = (k^{+} - k_{-})l/2$ between $(+)$ and $(-)$ circularly polarized modes in tin at $4.2\,\mathrm{K}$ as a function of $H$ for wave propagation along [001]. (After Fig. 6 in [23].)

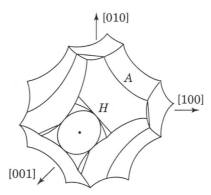

FIGURE 6.9. Fermi surface of aluminum in the second Brillouin zone. Line $A$ shows the belt of effective carriers for $\mathbf{k} \parallel \mathbf{H} \parallel [001]$ according to [24], Line $H$ according to [151].

[001], [110], and [111]. They reported results similar to those of Miller for $\mathbf{H} \parallel C_4$ relating to rotation, but found that the plot of harmonics for the single observed series did pass through the origin, in disagreement with reference [23]. The value of $(\partial S/\partial \kappa_z)/2\pi$ was found to be $0.58 \pm 0.04 \, \text{Å}^{-1}$, taking into consideration the selection rule $n = 1, 3, 5, \ldots$ for the four-fold symmetry in the direction of the field. The authors proposed that this resonance was caused by the hole-type carriers portrayed in Fig. 6.9.

More accurate DSCR measurements of the ultrasonic parameters for this metal were reported in [151]. The method used there was described in Sec. 2.2. The specimen had RRR $\approx 10^4$, the frequencies used were 42.8, 55.2, 62.4, and 74.6 MHz, and the temperature was 1.8 K. Every absorption peak showed the $N$-shaped variation in both polarization rotation as well as in ultrasonic velocity (see Fig. 6.10) expected for the deformation mechanism of interaction of ultrasound with conduction electrons. In addition, it was found that the DSCR harmonics followed the sequence $n = 1, 2, 3, \ldots$, which is forbidden for an axis with fourth-order rotational symmetry. This discrepancy was explained by the fact that the symmetry of the orbit can be lower than the symmetry of the entire FS cross section, as shown in Fig. 6.9. For the effective zone the derivative $|\partial S/\partial \kappa_z|$ was found to be $1.93 \pm 0.06 \, \text{Å}^{-1}$.

Jericho and Simpson [21] investigated attenuation and rotation of the polarization in copper using magnetic fields up to 45 kOe. Propagation was along the three principal crystallographic directions and the frequency range was 30 to 100 MHz. The specimens used with $\mathbf{k} \parallel \mathbf{H} \parallel [001]$ had RRR $= 390 \pm 10$ (Cu$_5$); 11 200 (Cu$_{1A}$); 19 700 (Cu$_1$).

The results obtained for the low-purity sample (Cu$_5$) showed good agreement with calculations based on the free-electron theory (see Fig. 6.11). In this case the parameter $k_0\ell$ was about 3.5 at 100 MHz.

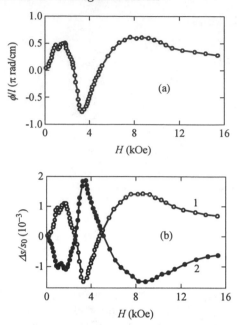

FIGURE 6.10. (a) Rotation of the polarization and (b) fractional change in velocity of circularly polarized waves [Curve 1 (−) polarization, Curve 2 (+) polarization] as functions of magnetic field in Al with [001] ∥ **k** ∥ **H**, $f = 74.6$ MHz, path length $l = 2.44$ mm, $T = 1.8$ K. (After Fig. 2 in [151].)

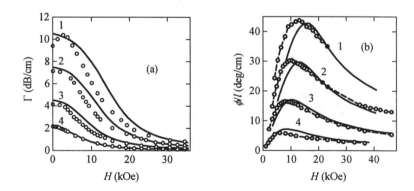

FIGURE 6.11. (a) Attenuation and (b) rotation of the polarization of ultrasound as functions of magnetic field for $Cu_5$ with **k** ∥ **H** ∥ [001], and $f = 89.9, 70.2, 50.2$, and $30.5$ MHz for $1-4$, respectively. Solid curves show calculations from the free-electron theory with $\tau/m_c = 0.99 \times 10^{16}$ s/g. (After Figs. 2 and 3 in [21].)

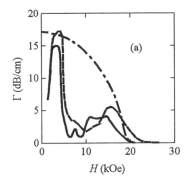 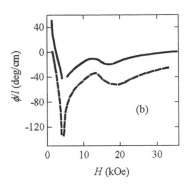

FIGURE 6.12. (a) Attenuation and (b) rotation of the polarization of ultrasound as functions of magnetic field in high-purity copper with $\mathbf{k} \parallel \mathbf{H} \parallel [001]$. (After Figs. 5 and 6 in [21].)

The situation was completely different for the higher-purity sample $Cu_1$ for which $k_0\ell$ was estimated to be 135 at 100 MHz. The results in rather low fields repeated the data by Boyd and Gavenda [20]. However in higher fields three additional peaks in attenuation and non-monotonic $H$-dependence of rotation angle up to 19 kOe were found. Since the absorption edge measures the Gaussian curvature at the maximum of $|\partial S/\partial \kappa_z|$, the magnitude of the derivative was estimated to be $1.4 \times 10^8$ cm$^{-1}$. Figure 6.12 shows the experimental data and the results of free-electron calculations for a sphere with radius $1.47 \times 10^8$ cm$^{-1}$ and a relaxation time corresponding to that derived from the resistivity ratio.

The observation in $Cu_1$ and $Cu_{1A}$ of absorption edges and in particular the position of peak 4 in the attenuation and the existence of a peak in the rotation angle for fields near $H_K$ supported calculations by Alig and Rodriguez [152]. However, the relative magnitudes of attenuation peaks 1 and 4 were in contradiction to their calculations.

Another metal, potassium, was studied by Blaney [18]. He carried out magnetoacoustic experiments in the frequency range 50 to 90 MHz with magnetic fields up to 60 kOe applied perpendicular and parallel to wave propagation at 4 K. The specimens were prepared of material with RRR = 3 000 and had $k_0\ell$ estimated to be 17 at 48.9 MHz. Rotation of the polarization (shown in Fig. 6.13) and attenuation were measured in a longitudinal field.

The effect on the magnetoacoustic properties of potassium of a possible spin-density-wave model proposed by Overhauser [153] for its ground state have been calculated by Alig et al. [154, 155]. Within experimental errors, the evidence agrees with the free-electron model, phonon–helicon interaction effects having been taken into account, and does not support the spin-density-wave model.

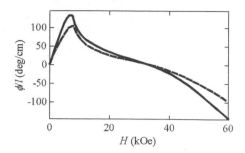

FIGURE 6.13. Rotation of the polarization of shear waves in potassium with
**k** ∥ **H** ∥ [001] at 48.4 MHz. The solid curve shows the measurements, while the
dashed curve shows the calculations for $k_0\ell = 15$. (After Fig. 7 in [18].)

Magnetoacoustic investigations of DSCR in tungsten were carried out by
Hui and Rayne [25]. The main purpose of the work was to use one more
effect (DSCR) to detail the FS of tungsten constructed on the basis of the
parametrized model proposed by Girvan, Gold, and Phillips (GGP) [156].

Although the FS of tungsten (Fig. 6.14) differs from a sphere, the elastic
properties are isotropic, thus it is possible to observe polarization phenomena even when **H** and **k** are directed in an arbitrary crystallographic
direction. In fact, in addition to large rotations for **H** ∥ **k** ∥ [001] (see
Fig. 6.15), the authors observed small differences in the signal dependence
on $H$ upon reversing the direction of the magnetic field for **H** ∥ **k** ∥ [110].

The experiments were carried out at 30 to 330 MHz using specimens having RRR = 80 000. Taking proper account of the rotation of the polarization
according to the method in [20] made it possible to measure precisely the
value of $|\partial S/\partial \kappa_z|$ and to determine the type of charge carriers responsible
for the principal absorption peak by using the sign of the angle of rotation
of the polarization angle in the vicinity of the peak. The revised FS using
the DSCR investigations from [25] is shown in Fig. 6.16.

It should be recalled that the relations

$$\Delta k^+(H) = \Delta k^-(H) \quad \text{and} \quad \Delta s^+(H) = \Delta s^-(H) \tag{6.4}$$

are valid for DSCR only if frequency dispersion can be neglected, i.e., $\omega$ can
be neglected in the resonant denominator $\nu - i\,(\mathbf{k}\cdot\bar{\mathbf{v}} - \omega - n\Omega)$ in (4.72). In
ultrapure metals $\omega$ can be comparable with or even exceed the relaxation
frequency $\nu$. Thus the resonances for $(+)$ and $(-)$ polarization occur at
slightly different fields, causing waves that were initially linearly polarized
to have ellipticity. Such experimental data can be properly treated with the
use of the method in [35]. After this paper appeared, several investigations
of the ellipticity of ultrasound under DSCR were published.

Vlasov et al. [157] studied tungsten (two specimens having RRR = 23 000
and 125 000) and molybdenum (RRR = 29 000), both for the geometry
**k** ∥ **H** ∥ [001] at 4.2 K in the frequency range 30 to 180 MHz. Two res-

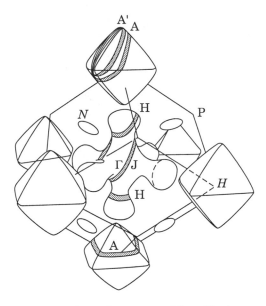

FIGURE 6.14. The Fermi surface of tungsten. The effective zones labeled are those having extremal area derivatives. Certain ellipsoids are omitted for clarity. (After Fig. 4 in [25].)

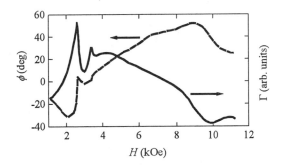

FIGURE 6.15. Attenuation (solid line) and rotation of the polarization (dashed line) of ultrasound as functions of magnetic field in tungsten with $\mathbf{k} \parallel \mathbf{H} \parallel [001]$ at 146.5 MHz. (After Fig. 2 in [25].)

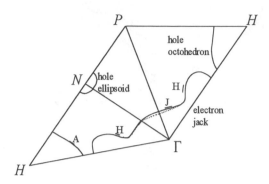

FIGURE 6.16. The central cross-sectional area of the Fermi surface of tungsten through the {001} and {1$\bar{1}$0} planes showing the difference between the GGP model (dashed lines) and the revised model (solid lines). The effective zones are labeled correspondingly. (After Fig. 6 in [25].)

onances (labeled $H$ and $B$ in Fig. 6.17) were considered. Both of them appeared as peaks in the absorption curves and therefore can be regarded as resulting from DSCR due to the deformation mechanism of interaction of ultrasound with conduction electrons. The $H$ resonance is due to electrons in the vicinity of the $\partial S/\partial \kappa_z$ maximum on the jack spheroid (see Fig. 6.14), whereas the origin of the $B$ resonance is not so clear. It is supposed to be due to the holes at the limiting point of the octahedron [30, 158] or to a line of inflection near this point [159].

In tungsten the $H$ resonance is stronger than the $B$, and vice versa in molybdenum. Figure 6.17 shows typical experimental curves obtained for tungsten. In order to fit the shape of the curves relating to the $H$ resonance, an analytically defined model of the FS, shown in Fig. 6.18, was introduced. The deformation potential was taken to be $\Lambda_{32} = Cp_y v_z$. The results of the fit for tungsten are presented in Fig. 6.19.

For molybdenum a good fit was obtained with $C = 3.7$ and $\tau = 0.1 \times 10^{-9}$ s. In addition, the Gaussian curvature $R_G = (1/2\pi)\partial S/\partial \kappa_z$ (in Å$^{-1}$) of the effective zones and the cyclotron masses of the effective electrons (in units of the free-electron mass $m_0$) were measured, either for the first time or else more accurately than previously (compare the data of [25, 30, 158, 159]). The results were: $R_G = 0.121 \pm 0.005$ and $0.162 \pm 0.006$, $m_c/m_0 = 0.35 \pm 0.11$ and $0.50 \pm 0.15$ (in tungsten for the $H$ and $B$ resonances, respectively); $R_G = 0.092 \pm 0.008$ and $0.173 \pm 0.008$, $m_c/m_0 = 0.21 \pm 0.06$ and $0.56 \pm 0.06$ (in molybdenum for the $H$ and $B$ resonances, respectively).

It should be pointed out that not only was ellipticity under DSCR discovered in [157] but also an attempt was made to measure the deformation potential parameter for the first time.

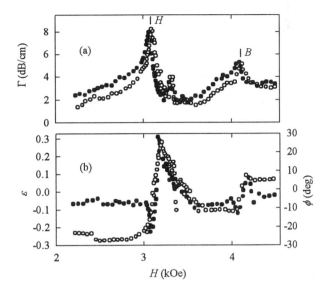

FIGURE 6.17. (a) Absorption of circularly polarized waves [filled circles (+) polarization, open circles (−)], and (b) ellipticity (filled circles) and rotation of the polarization (open circles) in tungsten with RRR = 125 000 as functions of magnetic field for $f = 175.5\,\mathrm{MHz}$, $L = 4.6\,\mathrm{mm}$. (After Fig. 1 in [157].)

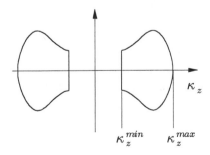

FIGURE 6.18. Analytical Fermi surface model for tungsten and molybdenum in the vicinity of the effective zone related to the $H$ resonance. (After Fig. 6 in [157].)

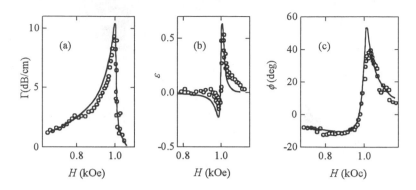

FIGURE 6.19. Line shapes of (a) absorption, (b) ellipticity, and (c) rotation of the polarization for the $H$ DSCR in tungsten. Open circles: experimental values at 174.4 MHz, $l = 1.4$ cm, RRR = 125 000. Solid line: theoretical fit with $C = 3.9$, $\tau = 0.6 \times 10^{-9}$ s, $X = \Omega/k_0 v_{\text{zres}}$. (After Fig. 8 in [157].)

Later, investigation of ellipticity in the vicinity of sharp peaks of ultrasonic absorption was suggested as a method for measuring relaxation time. Rinkevich et al. [160] obtained the following formula, valid for $\omega\tau \ll 1$:

$$\tau = \frac{nc}{H_{max} - H_0} \left| \frac{m_c}{e} \right| \left[ \left( 1 - \frac{2}{\sqrt{5}} \right)^{1/2} + \frac{1}{\sqrt{3}} \right], \tag{6.5}$$

where $H_{max}$ is the field corresponding to the maximum of ellipticity, $H_0$ to the zero between the maximum and minimum.

Both odd and even contributions to magnetic field dependences of $\varepsilon$ and $\phi$ were determined by Vlasov and Rinkevich [161] in tungsten for ultrasound propagating along [110] $\parallel$ **H**. In the case of wave propagation along the two-fold symmetry axis, polarization of the normal modes is elliptical in general and depends on the components of the effective elastic moduli, some of them being odd and some even functions of $H$. It can be seen in Fig. 6.20 that odd contributions to $\phi$ versus $H$ dominate for initial polarization $\mathbf{U}_0 \parallel [001]$ while even contributions dominate when $\mathbf{U}_0$ makes an angle $\pi/4$ with [001] in the (110) plane. As for $\varepsilon(H)$, it is obvious that odd as well as even contributions are present for both initial conditions.

## 6.1.1  Anomalous Sound Propagation

Finally, in concluding the discussion of DSCR one more effect should be mentioned. It occurs when the mean free path $\ell$ of the conduction electrons is comparable to or greater than the spatial extent $\Delta z$ of the ultrasonic wave packet propagating through a specimen. Obviously, it can be observed only in experiments employing the pulse technique. For an ultrasonic pulse duration $T = 1\,\mu$s and sound velocity $s = 3 \times 10^5$ cm/s, one must have

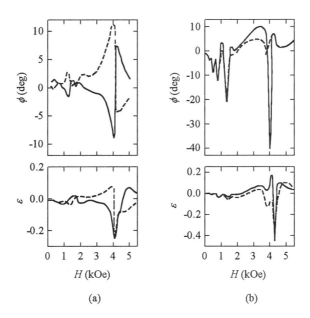

FIGURE 6.20. Rotation of the polarization and ellipticity versus magnetic field in tungsten at 159 MHz with RRR = 102 000, $T = 4.2$ K, and $\mathbf{k} \parallel [110]$. (a) Initial polarization $\mathbf{U}_0 \parallel [001]$. (b) $\mathbf{U}_0$ at an angle of $\pi/4$ from [001] in the (110) plane. Solid lines are for $\mathbf{H} \parallel \mathbf{k}$; dashed lines are for $\mathbf{H} \parallel -\mathbf{k}$. (After Figs. 5 and 6 in [161].)

$\ell = v_F \tau \sim \Delta z = sT \sim 1.5\,\text{mm}$ to observe this effect. Another way of expressing this condition is that the electron relaxation time $\tau$ must be of the order of the transit time for the electrons through the wave packet, $\Delta t = \Delta z / v_F = (s/v_F)T \sim (3 \times 10^5\,\text{cm/s})/(10^8\,\text{cm/s}) \times 5 \times 10^{-7}\,\text{s} = 1.5 \times 10^{-9}\,\text{s}$.

Although most of the ultrasonic experiments described here use the pulse technique, the typical electron transit times $\Delta t$ are much larger than the relaxation times of the conduction electrons. Therefore, a gated detector set to measure the voltage in the middle of the ultrasonic pulse may be regarded as a device that determines the amplitude of a continuous harmonic wave with frequency equal to that of the dominant component of the harmonics which make up the wave packet. In this case one observes that, while the height of the pulse on the oscilloscope screen varies as the magnetic field is swept, its waveform remains unchanged.

The situation is quite different when the relaxation time is of the same order as, or exceeds, the transit time $\Delta t$. In this case nonequilibrium electrons (i.e., electrons excited by the lattice displacements and the electromagnetic fields accompanying these shifts) spread out of the main body of the wave packet (we will call it the normal packet since it propagates with the ordinary sound velocity) and interact with the lattice to produce an ultrasonic wave field outside the normal pulse. As a result the initial form of the pulse is distorted to a degree that depends on the strength of the interaction between ultrasound and the conduction electrons.

This phenomenon is called anomalous propagation since the ultrasonic field spreads with the Fermi velocity. In both DSCR and the Faraday effect it can be observed that the polarization of the ultrasonic field transported by the electrons (outside the normal packet) is quite different from that inside. Thus anomalous propagation of ultrasound has some specific implications for polarization phenomena.

Anomalous propagation of ultrasound was first observed in ultra-pure gallium by Fil et al. in connection with magnetoacoustic geometric oscillations (sometimes referred to as Pippard resonances) [162] and with DSCR [163]. Gavenda and Casteel observed anomalous propagation of ultrasound in copper in connection with open-orbit resonances [164] as well as for DSCR [165]. In all of these experiments the ultrasound traveled along an axis having two-fold symmetry so that the polarization of the waves could not be changed appreciably by the magnetic field.

Gudkov and Fil carried out an experiment in indium [166] with the direction of both sound propagation and magnetic field parallel to the tetragonal axis. Thus electrons (more properly, holes, in the case of indium) moving through the normal wave packet absorbed ultrasonic energy and then, leaving the packet because of their higher velocity, spread it coherently in the region outside the packet. Since, for DSCR and $\mathbf{H} \parallel C_4$, one circularly polarized normal mode interacts with the electrons drifting along $\mathbf{H}$ and the other with the electrons drifting in the opposite direction, the structures of the ultrasonic field in front of and behind the normal packet correspond

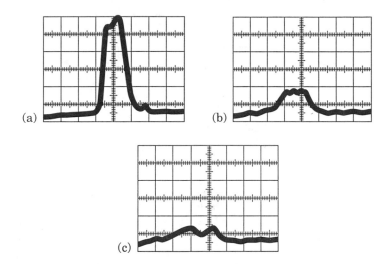

FIGURE 6.21. Oscillograms of the detector output signal for 65.2 MHz shear waves in indium with RRR = 80 000 at $T = 1.75$ K (**k** $\parallel$ **H** $\parallel$ [001]). The horizontal scale is $0.4\,\mu s$/div. (a) Absence of anomalous sound propagation (ASP) ($H = 8.20$ kOe). (b) ASP present for DSCR at $H = 7.45$ kOe, and (c) 7.40 kOe. (After Fig. 6 in [166].)

to one or the other polarizations. With DSCR the most effective transport of the ultrasonic field occurs when peaks of ultrasonic absorption caused by the deformation mechanism [150] are observed. Typical oscillograms obtained in these experiments are shown in Fig. 6.21.

In [165] it was noted that gating the signal before the arrival of the normal packet gives an opportunity to unravel the complex features of DSCR experiments. This is due to the fact that timing the gating signal in this fashion allows detection of the signal associated with the field transported by the electrons while rejecting the signal from the normal wave packet. Consequently, an output signal from the detector is present only when DSCR occurs and other factors which cause variations in the ultrasonic signal are absent.

Under the conditions for the Faraday effect a new and unique feature arises: the experimentalist need not pay attention to the polarization direction of the receiving transducer since the anomalously propagated wave will be circularly polarized. In addition, and what is even more important, it is possible to determine the resonant magnetic field on the basis of direct observation of signals rather than having to perform the rather complicated calculations of $\Gamma^{\pm}(H)$ described in Chapter 2. Thus in [166] the Gaussian curvature of the resonant electron band $[R_G = (0.20 \pm 0.001)\,\text{Å}^{-1}]$ was determined using the plot given in Fig. 6.22.

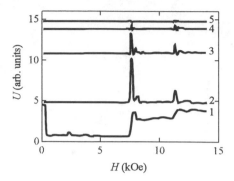

FIGURE 6.22. Magnetic field dependence of the gated detector output signal for indium under the same conditions as in Fig. 6.21, except that the frequency is 62.8 MHz. Curve 1 shows the amplitude of the normal wave packet. Curves 2 – 5 have been obtained for gating times $\tau_g = 0.0$, 0.2, 0.4, and $0.6\,\mu$s, respectively, before the expected arrival of the normal wave packet, and are shifted vertically by various amounts for clarity. (After Fig. 3 in [166].)

## 6.2    Magnetoacoustic Geometric Oscillations

The first magnetoacoustic phenomena discovered (and the ones most widely studied) were the so-called geometric oscillations (GO). The first hint that such effects were possible was a report by Bömmel in 1955 of a minimum in the attenuation of ultrasound as a function of magnetic field in tin [167]. Morse suggested that one might expect oscillatory variations in ultrasonic attenuation for $\mathbf{k} \perp \mathbf{H}$ due to resonant absorption when the Fermi electron cyclotron orbit diameter is an integral multiple of the ultrasonic wavelength $\lambda$ [168]. For a general Fermi surface, the extremal dimension of the cyclotron orbit in a plane perpendicular to the magnetic field is given by $D = p_\perp/eH$, where $p_\perp$ is the component of electron momentum perpendicular to both $\mathbf{H}$ and $\mathbf{k}$ and $e$ is the electronic charge, thus resonances might be expected for $n\lambda = p_\perp/eH$. A plot of ultrasonic absorption versus $1/H$ should have oscillations with a period given by

$$\Delta(1/H) = e\lambda/p_\perp. \tag{6.6}$$

Morse and Bohm found a series of absorption maxima and minima in polycrystalline indium that could be interpreted in this fashion [169]. Subsequent measurements in polycrystalline copper were in good agreement with the known electronic properties of that metal [170]. Full confirmation of the spatial resonance model was delayed by the lack of single-crystal specimens in which the electron mean free path exceeded the ultrasonic wavelength, but in 1959 Morse and Gavenda reported seven maxima and minima in attenuation in a copper crystal [7]. When plotted against inverse field, the oscillations exhibited the periodicity expected for electrons having the Fermi energy in copper. Furthermore, the data showed anisotropy that directly correlated with crystal direction.

GO in both the absorption and the velocity of ultrasound in a transverse magnetic field have been widely used for mapping the Fermi surfaces of metals. To learn more about this effect see the reviews by Roberts [70] and Mertsching [71] and related chapters in the books by Truell, Elbaum, and Chick [72] and by Tucker and Rampton [73].

Here we will discuss a less familiar variant of magnetoacoustic GO, namely those that occur in a longitudinal magnetic field, which, in conjunction with the frequency dispersion of the medium, may result in oscillations of both the ellipticity and the rotation of the polarization. These effects have been observed in indium and interpreted by Gudkov and Tkach [67, 171, 172].

The first observation of oscillations of ultrasonic absorption for $\mathbf{k} \parallel \mathbf{H}$ was made by Rayne and Chandrasekhar [173] in indium. The idea that such oscillations could exist was first proposed by Mackinnon et al. [174]. The authors presented a qualitative physical model in which ultrasonic absorption could exhibit oscillations, but applied their model to the absorption peaks in cadmium that appear to have a quite different origin, namely DSCR. In addition, Daniel and Mackinnon [175] attempted to obtain an expression connecting the period of the oscillations with characteristics of the FS; however, what they derived were the conditions for DSCR peaks.

By using a quantum-mechanical model to treat the problem of propagation of longitudinal ultrasonic waves along $\mathbf{H}$ and a model of the FS in the shape of ellipsoid (one of its axes inclined at an angle to $\mathbf{H}$), Quinn [176, 177] showed that in the collisionless limit the corresponding component of the conductivity tensor contained an oscillatory term. His principal result was a correct expression for the period of the GO. The same problem but in the frame of a simpler quasiclassical theory was solved by Eckstein [178] who obtained an analogous expression. Later experiments in antimony by Backman et al. [179], Eckstein et al. [180], and Miller [23] confirmed the correctness of the Quinn–Eckstein expression for the period of oscillations. However, to observe these oscillations a more stringent (by a factor of $v_F/s \sim 10^3$) condition,

$$\omega \tau \gtrsim 1, \qquad (6.7)$$

is required in comparison with that for $\mathbf{k} \perp \mathbf{H}$:

$$k\ell \gtrsim 1. \qquad (6.8)$$

The important role of frequency dispersion for this effect was outlined in [67]. Here we will discuss the experiments performed in indium and a quasiclassical theory as given in [171].

Let us consider the solutions of the dispersion equation (4.77) in the weak-coupling approximation. For the elastic-like modes they were given by equation (4.92), which we will rewrite as

$$k^\pm = k_0 \left[1 - \frac{i\omega\alpha^\pm + H^2/4\pi}{2c_{1313}} - \frac{1}{2c_{1313}} \frac{\left(\omega\beta^\pm \pm icHk_0/4\pi\right)^2}{c^2k_0^2/4\pi + i\omega\sigma^\pm}\right]. \qquad (6.9)$$

Remember that equation (4.92) was derived for wave propagation along a four-fold rotation symmetry axis (i.e., $k^\pm \parallel H \parallel C_4$).

Any electroacoustic coefficient $\gamma^\pm$ [i.e., $\alpha^\pm$, $\sigma^\pm$, and $\sigma^+$ defined by equations (4.80)–(4.81)] may be written as

$$\gamma^\pm = \frac{2\pi}{h^3} \int dp_z \sum_\kappa |m_c|(\mathbf{a}, \mathbf{b})^\pm, \qquad (6.10)$$

where $\mathbf{a}$ and $\mathbf{b}$ are two-dimensional vectors lying in the $p_x$ - $p_y$ plane and the index $\kappa$ numbers the orbits existing at a given $p_z$ value. The vectors $\mathbf{a}$ and $\mathbf{b}$ stand for either $e\mathbf{v}_\perp = e(v_1, v_2)$ or $\mathbf{w}_\perp = (\Lambda_{13}, \Lambda_{23})$. To calculate the electroacoustic coefficients one has to assume

$$\text{for } \alpha^\pm: \quad \mathbf{a} = \mathbf{b} = \mathbf{w}_\perp; \qquad \text{for } \beta^\pm: \quad \mathbf{a} = \mathbf{w}_\perp, \quad \mathbf{b} = e\mathbf{v}_\perp;$$

$$\text{for } \sigma^\pm: \quad \mathbf{a} = \mathbf{b} = e\mathbf{v}_\perp. \qquad (6.11)$$

The expression in brackets on the right-hand side of equation (6.10) is written as

$$(\mathbf{a}, \mathbf{b})^\pm = \sum_{n=-\infty}^{\infty} \frac{a_n^\pm b_n^\pm}{\nu - i\left(k_0\bar{v}_z - \omega - n\Omega\right)}. \qquad (6.12)$$

The factors in the numerator of equation (6.12) are calculated by means of the definitions

$$a_n^\pm \equiv \frac{1}{2\pi} \int_0^{2\pi} d\Theta \, |\mathbf{a}_\perp| \exp(\pm i\varphi_a) \exp\{-i[k_0 Z(\Theta) + n\Theta]\},$$

$$b_n^\pm \equiv \frac{1}{2\pi} \int_0^{2\pi} d\Theta \, |\mathbf{b}_\perp| \exp(\mp i\varphi_b) \exp\{i[k_0 Z(\Theta) + n\Theta]\}, \qquad (6.13)$$

where $\varphi_a$ and $\varphi_b$ give the directions of $\mathbf{a}_\perp$ and $\mathbf{b}_\perp$ in the $p_x$ - $p_y$ plane, and

$$Z(\Theta) = \frac{1}{\Omega} \int_0^{\Theta'} d\Theta \, (v_z - \bar{v}_z). \qquad (6.14)$$

Recall that equation (4.58) defines $\Theta = \Omega t$ as the dimensionless time for an electron traveling along the cyclotron trajectory, therefore $Z(\Theta) \sim 1/H$ is the distance covered by an electron in the $H$ direction during the interval $(0, t')$ measured in the coordinate system moving along $z$ with the velocity $\bar{v}_z$.

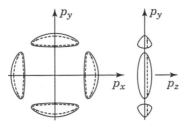

FIGURE 6.23. Model of a multiply connected Fermi surface with the cross sections indicated by dashed lines corresponding to the condition $\bar{v}_z = s_0$. Points 1 and 2 designate turning points that are determined in the coordinate system moving with the ultrasonic wave. (After Fig. 3 in [67].)

As can be seen in equations (6.10)–(6.12), an electroacoustic coefficient is the sum of a series whose terms are numbered by the index $n$. For any $n \neq 0$, the denominator in equation (6.12) has a resonant character related to DSCR. The term with $n = 0$ has quite a different nature. This term is not resonant with respect to the magnetic field; it describes cyclotron absorption of ultrasound in a strong (above the Kjeldaas edge) magnetic field (see, for example, [181, 182] and references given there). As will be seen later, it is just this term that describes GO.

Let us consider in detail the contribution of the $n = 0$ term. One obtains from (6.13) and (6.14)

$$(\mathbf{a}, \mathbf{b})_{m=0}^{\pm} = \frac{a_0^{\pm} b_0^{\pm}}{\nu - i\left(k_0 \bar{v}_z - s_0\right)}, \tag{6.15}$$

$$a_0^{\pm} = \frac{1}{2\pi} \int_0^{2\pi} d\Theta \, |\mathbf{a}_\perp| \exp(\pm i\varphi_a) \exp\{-i[k_0 Z(\Theta)]\},$$

$$b_0^{\pm} = \frac{1}{2\pi} \int_0^{2\pi} d\Theta \, |\mathbf{b}_\perp| \exp(\mp i\varphi_b) \exp\{i[k_0 Z(\Theta)]\}. \tag{6.16}$$

In accordance with Kotkin's theorem [183], orbits having an $r$-fold symmetry axis $C_r$ contribute to the sum (6.12) for $n = ru \pm 1$ (where $u = 0, \pm 1, \pm 2, \dots$). This is why GO may be determined only by electrons on orbits without any rotational symmetry (i.e., under $r = 1$). Such orbits appear for highly symmetric directions if the FS cross section by a plane $p_z = \text{const}$ is multiply connected. These types of orbits are shown in Fig. 6.23. Specifically, they exist on the $\alpha$ and $\beta$ arms in aluminum or on $\beta$ arms in indium. The FSs of these metals may be seen in [91].

From (6.14) it follows that $Z(\Theta)$ has a period of $2\pi$. Hence, the number of minima for this function on an orbit in the interval $0 \leq \Theta \leq 2\pi$ always equals the number of maxima. Hereafter these extremal points on the orbit will be referred to as *turning points*. Actually, they are the electron's turning points in a coordinate system moving with velocity $\bar{v}_z$ in the $\mathbf{k}$ direction. There will always be at least two turning points on an orbit [with the exception of the trivial case $v_z(\Theta) \equiv \bar{v}_z$]. Let us consider this simplest situation.

For $k_0 Z_{max} \gg 2\pi$ the expressions (6.16) represent integrals of a rapidly oscillating function of the argument $\Theta$. If the turning points are numbered such that $\Theta_1$ corresponds to the maximum of $Z(\Theta)$ and $\Theta_2$ to the minimum, then the following may be obtained using the stationary phase method (see for example [184]):

$$a_0^{\pm} = \sqrt{\frac{|\Omega|}{2\pi k_0}} \sum_{m=1}^{2} \frac{|\mathbf{a}_{\perp}(\Theta_m)|}{\sqrt{|\partial v_z/\partial \Theta|_m}} \exp[-i\Phi_a^{\pm}(\Theta_m)],$$

$$b_0^{\pm} = \sqrt{\frac{|\Omega|}{2\pi k_0}} \sum_{m=1}^{2} \frac{|\mathbf{b}_{\perp}(\Theta_m)|}{\sqrt{|\partial v_z/\partial \Theta|_m}} \exp[i\Phi_b^{\pm}(\Theta_m)],$$

(6.17)

where

$$\Phi_c^{\pm}(\Theta_m) = k_0 Z(\Theta_m) + (-1)^m \pi/4 \mp \varphi_c,$$ (6.18)

and the index $c$ stands for $a$ or $b$. Let us restrict our treatment to the consideration of only those orbits that have a mirror symmetry axis. Then the vectors $\mathbf{c}_{\perp}$ ($\mathbf{c}_{\perp} = \mathbf{a}_{\perp}$ or $\mathbf{b}_{\perp}$) at the turning points $r = 1, 2$ become connected by the symmetry relations

$$|\mathbf{c}_{\perp}(\Theta_1)| = |\mathbf{c}_{\perp}(\Theta_2)|, \qquad \varphi_c(\Theta_1) = -\varphi_c(\Theta_2).$$ (6.19)

As a result, the numerator of (6.15) takes the form

$$a_0^{\pm} b_0^{\pm}$$

$$= \frac{|\Omega|}{2\pi k_0} \sum_{p=1}^{2} \sum_{m=1}^{2} \frac{|\mathbf{a}_{\perp}(\Theta_1)||\mathbf{b}_{\perp}(\Theta_1)|}{|\partial v_z/\partial \Theta|_1} \exp\left\{ i\left[ \Phi_b^{\pm}(\Theta_p) - \Phi_a^{\pm}(\Theta_m) \right] \right\}.$$ (6.20)

The indices $p$ and $m$ denote the turning points on the cyclotron orbit. Of the four terms in the right-hand part of equation (6.20), two correspond to the condition $p = m$. The contribution of this pair to the sum is

$$(a_0^{\pm} b_0^{\pm})_{p=m} = \frac{|\Omega|}{2\pi k_0} \frac{|\mathbf{a}_{\perp}(\Theta_1)||\mathbf{b}_{\perp}(\Theta_1)|}{|\partial v_z/\partial \Theta|_1} \cos\left[ \varphi_b^{\pm}(\Theta_1) - \varphi_a^{\pm}(\Theta_1) \right].$$ (6.21)

The sum of the other two terms is of greater interest for us:

$$(a_0^{\pm} b_0^{\pm})_{p \neq m} = \frac{|\Omega|}{2\pi k_0} \frac{|\mathbf{a}_{\perp}(\Theta_1)||\mathbf{b}_{\perp}(\Theta_1)|}{|\partial v_z/\partial \Theta|_1} \cos\left( k_0 \Delta Z \mp \Delta \varphi_{ba}^{\pm} - \frac{\pi}{2} \right),$$ (6.22)

where $\Delta Z = Z(\Theta_1) - Z(\Theta_2)$ is the dimension of the cyclotron orbit along the $z$ axis measured in the coordinate system moving with velocity $\bar{v}_z$, and $\Delta \varphi_{ba} = \varphi_b(\Theta_1) - \varphi_a(\Theta_2)$ is the phase shift characteristic of the given orbit.

To sum up,

$$\omega\gamma_0^{\pm} = \frac{2|e|H}{ch^3 k_0} \int dp_z \sum_{\kappa} \frac{|\mathbf{a}_{\perp}(\Theta_1)||\mathbf{b}_{\perp}(\Theta_1)|}{|\partial v_z/\partial\Theta|_1}$$

$$\times \left\{\cos\left[\varphi_b(\Theta_1) - \varphi_a(\Theta_1)\right] + \sin\left(k_0\Delta Z \mp \Delta\varphi_{ba}\right)\right\} \frac{\omega\tau}{1 - i\omega\tau\left(\bar{v}_z/s_0 - 1\right)}.$$

$$(6.23)$$

We have given the electroacoustic coefficients multiplied by $\omega$ since they occur in that form in equation (6.9).

According to (6.14), $\Delta Z \sim 1/H$. As a result, the integral in (6.23) contains an oscillating function (with respect to the reciprocal of the magnetic field) and the oscillation phases for the (+) and (−) components differ by $2\Delta\varphi_{ba}$. The parameter $\omega\tau$ characterizing the frequency dispersion is of great importance for the integral over $p_z$. For sufficiently large values of $\omega\tau$, a narrow belt on the FS is selected. It contains those electrons that interact most effectively with the ultrasonic wave. On the effective orbits $\bar{v} \simeq s_0$, and the belt's width $\Delta\bar{v}_z$ equals $s_0(\omega\tau)^{-1}$. In the collisionless limit $(\omega\tau \to \infty)$ the real part of the last factor on the right in equation (6.23) coincides (within a constant) with the $\delta$ function; the integration will turn out to yield a single oscillatory function, determined for the orbit $\bar{v}_z = s_0$. As the parameter $\omega\tau$ decreases, the effective orbit belt becomes wider and harmonic contributions, having comparable amplitudes but different phases, add up. This considerably decreases the oscillatory component of $\gamma^{\pm}$.

The above may be formulated in the following way. According to equation (6.23), the oscillatory contributions to the electroacoustic coefficients contain a geometric factor $\sin(k_0\Delta Z \mp \Delta\varphi_{ba})$ with the argument $k_0\Delta Z \sim k_0/H$ being the spatial dispersion parameter. At the same time, the effective orbits are determined by another parameter, namely $\omega\tau$. Therefore, in the general case the result of the integration in equation (6.23) is determined by the two parameters, $\omega\tau$ and $k_0\Delta Z$, which arise from taking into account frequency and spatial dispersion, respectively.

To illustrate these conclusions, we will describe the GO caused by four sheets of a FS that coincide with one another under $\pi/4$ rotations (as shown in Fig. 6.23). At any reasonable value of $\omega\tau$ (from 0.1 to 10) only a rather narrow belt of orbits will be effective. Within its bounds one can assume that the values of $\mathbf{a}_{\perp}(\Theta_1)$, $\mathbf{b}_{\perp}(\Theta_1)$, $(\partial v_z/\partial\Theta)_1$, $[\varphi_b(\Theta_1) - \varphi_a(\Theta_1)]$, $\Delta\varphi_{ba}$, and $m_c$ are independent of $p_z$. Also, the range of integration may be extended to $\pm\infty$.

Hereafter we will assume (by analogy with the case of an elliptical FS) that within the effective belt $dp_z = M\,d\bar{v}_z$, where $M$ is a constant with the dimensionality of mass. Let us take a simple linear model for the $\Delta Z$ dependence on $\bar{v}_z$:

$$\Delta Z = \Delta Z_s\left[1 + \tilde{\chi}\left(\bar{v}_z/s_0 - 1\right)\right] \equiv \frac{\Delta G_s}{H}\left[1 + \frac{\chi}{\Delta G_s}\left(\bar{v}_z/s_0 - 1\right)\right]. \quad (6.24)$$

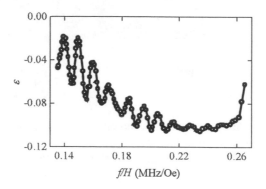

FIGURE 6.24. Plot of ellipticity for indium with [001] $\parallel \mathbf{k} \parallel \mathbf{H}$ as a function of $f/H$. $f = \omega/2\pi = 63\,\text{MHz}$, $T = 1.3\,\text{K}$, $l = 2\,\text{mm}$. (After Fig. 3 in [171].)

Here and below the subscript $s$ means that the given quantity is calculated over the cyclotron orbit with $\bar{v}_z = s_0$.

Performing the integration in (6.23), we obtain

$$\omega\gamma^{\pm}_{m=0} = B\frac{H}{k_0}\left\{ \cos\left[\varphi_b(\Theta_1) - \varphi_a(\Theta_1)\right] \right.$$
$$\left. + \exp\left(-\frac{k_0|\chi|}{H\omega\tau}\right)\left[\sin\psi^{\pm}_{ba} + i\cos\psi^{\pm}_{ba}\text{sgn}(\chi)\right]\right\}_s , \quad (6.25)$$

where the constant $B$ and the oscillation phase $(\psi^{\pm}_{ba})_s$ are given by

$$B = \left[\frac{8\pi s_0 M|e|}{h^3 c}\frac{|\mathbf{a}_{\perp}(\Theta_1)||\mathbf{b}_{\perp}(\Theta_1)|}{|\partial v_z/\partial\Theta|_1}\right]_s ,$$
$$(\psi^{\pm}_{ba})_s = k_0\Delta Z_s \mp (\Delta\varphi_{ba})_s. \quad (6.26)$$

Equation (6.25) contains functions that are oscillatory with respect to $H^{-1}$, the GO phases in $\gamma^+$ and $\gamma^-$ being different in the general case. The oscillation amplitude vanishes for $\omega\tau \to 0$ and approaches the value $BH/\omega k_0$ as $\omega\tau \to \infty$.

GO of ellipticity (see Fig. 6.24) have been observed in indium at temperatures below $3\,\text{K}$. The specimen had RRR $\gtrsim 6 \times 10^4$ and dimensions of $10 \times 10 \times 2\,\text{mm}^3$ with its tetragonal axis normal to the plate surface. The magnetic field and wave vector $\mathbf{k}$ were parallel to the [001] axis. The accuracy of orientation was approximately $4°$.

In the interval $0.13 < f/H < 0.2$ the oscillations of $\Gamma^+$ and $\Gamma^-$ were nearly sinusoidal and shifted with respect to one another by a half period as shown in Fig. 6.25.

Oscillations in $s^+$ and $s^-$ appear, also, but they are in phase and almost equal in amplitude. The differences in the amplitudes of the oscillations caused oscillations of the rotation of the polarization to be registered in

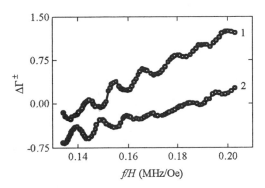

FIGURE 6.25. Absorption of circularly polarized waves, where $\Delta\Gamma^{\pm} = \Gamma^{\pm} - \Gamma_0$ and $\Gamma_0 = \Gamma(H = 0, T = 4.2\,\mathrm{K})$. Curves 1 and 2 are for $(+)$ and $(-)$ polarization, respectively; the remaining parameters as in Fig. 6.24. (After Fig. 2 in [171].)

higher magnetic fields and at higher frequencies where the oscillations of the ellipticity became very large (see Fig. 6.26).

Where could the effective electrons be located? It was mentioned above that orbits of lowered symmetry in indium exist on the $\beta$ arms of the FS in the third Brillouin zone. But if one calculates a typical $v_z$ variation for the effective orbit using the period of oscillations obtained from (6.21),

$$\Delta\left(\frac{1}{H}\right) = \left| \frac{2\pi e}{k_0 m_c} \left[ \int_{\Theta_1}^{\Theta_2} d\Theta \, (v_z - s_0) \right]^{-1} \right|, \qquad (6.27)$$

it can be proved that the $v_z$ variations are comparable with the Fermi velocity $v_F$, not with $s_0$, as would be the case for the precise orientation [001] $\|$ **H**. It was proposed that orbits with $v_z$ variations of the order of $v_F$ may appear for a small deviation of the vector **H** from the [001] direction as shown in Fig. 6.27. However, there is a problem with describing the normal modes as circularly polarized waves, which casts doubt on the applicability of the expressions given here.

From a formal point of view any deviation of the magnetic field from the [001] direction breaks the four-fold symmetry of the orbit; new nonzero components of the elastic moduli (of purely electronic origin) appear, resulting in the normal modes becoming elliptically polarized. But a small (few degrees) deviation from the tetragonal axis qualitatively (up to changing the connectivity) varies the cyclotron orbits on the $\beta$ arms only. At the same time, the main group of the current carriers (95%) belonging to the second Brillouin zone undergoes small variations of their orbits. As a result, the normal mode polarization differs only slightly from circular, allowing us to use the relations presented above.

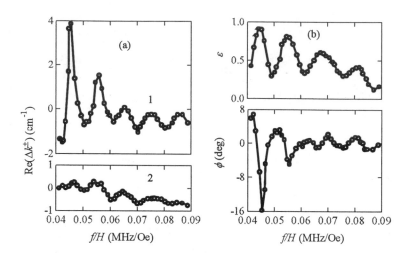

FIGURE 6.26. Dependences of (a) $\mathrm{Re}\left(\Delta k^{\pm}\right) \sim -\Delta s^{\pm}$ and (b) $\varepsilon$ and $\phi$ on $f/H$. $f = 76\,\mathrm{MHz}$, $T = 1.3\,\mathrm{K}$, and $l = 1.5\,\mathrm{mm}$. Curves 1 and 2 are for $(+)$ and $(-)$ polarization, respectively. (After Fig. 6 in [171].)

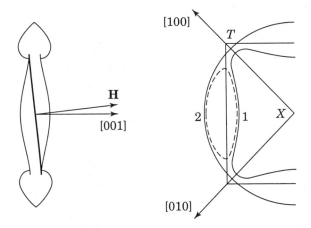

FIGURE 6.27. Effective orbit on the Fermi surface of indium in the third Brillouin zone for a deviation of **H** from [001] in (110) plane. The positions 1 and 2 designate the states conforming to the turning points. (After Fig. 7 in [171].)

Recall that there are two interactions with conduction electrons that determine the magnetic field dependences of absorption and phase velocity of ultrasonic waves: deformation and electric field. If equation (6.25) is written for $\alpha^\pm$, it turns out that the GO in absorption and ultrasonic velocity caused by the deformation interaction must have equal amplitudes and a phase shift of $2(\Delta\varphi_{ww})_s$ for $(+)$ and $(-)$ polarized waves.

Experimentally, the amplitudes for both the absorption and phase velocity are different for the different polarizations (particularly at $H > 800$ Oe for $f = 60 - 80$ MHz where the oscillations even exhibit distortion from their initially harmonic shape); as for the oscillation phases, these coincide for $s^+$ and $s^-$, but differ by $\pi$ for $\Gamma^+$ and $\Gamma^-$. These circumstances made it possible to deduce that the GO in indium are related to the electric field interaction.

Since the oscillatory parts of the electroacoustic coefficients contribute to the expression for the wave vector (6.9) in different ways (linearly for $\alpha^\pm$, nonlinearly for the others), it is difficult to guess what the form of the oscillations will be in this or that case. However, at low fields, where the amplitudes of the oscillations of different coefficients are small compared with their non-oscillatory components, it is possible to describe $\Delta k^\pm$ in a linear approximation for $\gamma^\pm_{m=0}$. All the terms in such an expression should have the same $\tau$-dependence given by equation (6.25). Therefore, an opportunity arose to study the temperature dependence of the relaxation time of effective charge carriers. In [171] it was found that the relaxation time in the interval 1.3 to 2.5 K obeys the law $\tau^{-1} = \tau_0^{-1} + aT^3$. This result supplemented experimental data presented in [185–187], which provided an opportunity to judge the nature of the electron–phonon scattering in indium.

## 6.3   Doppleron–Phonon Resonance

Tsymbal and Butenko [28] were the first to observe doppleron–phonon resonance (DPR) in an ultrasonic experiment. Their investigation in and those by Galkin et al. [30, 31] in molybdenum and tungsten did not take into account polarization phenomena, whereas Vlasov and Gudkov [36, 37] measured ellipticity and polarization rotation in tungsten and indium in the presence of DPR. The latter papers were the first applications of the method for measuring both $\varepsilon$ and $\phi$ [35]. Later, Golik et al. [188, 189] used a combined acousto-electromagnetic technique for investigating DPR in tungsten: ultrasonic waves were generated by an electromagnetic circuit and detected by a piezoelectric transducer or vice versa.

From the very beginning of DPR investigations it was necessary to distinguish between this phenomenon and DSCR. The principal distinguishing feature was a nonlinear dependence of the resonant field on the frequency

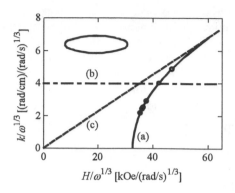

FIGURE 6.28. (a) Doppleron–phonon spectrum near DSCR for the electron lens in cadmium (the cross section of this sheet of the FS is shown in the upper left corner). The experimental points correspond to frequencies of 29, 35, 36.5, 46, 74, and 99 MHz. ((b) Transverse ultrasonic wave spectrum for $f = 74$ MHz. (c) Doppleron spectrum. (After Fig. 1 in [28].)

(see Fig. 6.28), in contrast with DSCR, which has a quite linear dependence. However, the magnitude of the resonant field, as can be seen in Fig. 6.29, was very difficult to determine without data on the propagation characteristics of circularly polarized waves. Even when dealing with circular components of the received signal it is necessary to consider which type of coupling is involved, weak or strong. The answer to this question will determine whether it is possible to obtain the absorption and phase velocity as functions of magnetic field for particular normal modes. The simplest way of determining the value of the resonant field (which is to find the field at which the circularly polarized elastic-like mode has maximum absorption) is possible only in the weak-coupling case. In addition, it is of particular interest to find whether the polarization rotation and ellipticity have appreciable values under DPR.

This problem was resolved in [36] while investigating ultrasonic propagation in tungsten. The nearly cylindrical specimen was 4.57 mm long and approximately 6 mm in diameter with RRR $\simeq$ 125 000. The ultrasound was in the frequency range of 54.3 to 181.3 MHz and propagated along the fourth-order symmetry axis. Figure 6.30 shows typical resonance curves obtained at the lowest frequency of the range studied. As can be seen, the values of $\varepsilon$ and $\phi$ measured at this frequency and propagation length $L$ were rather small. At 181.3 MHz and $l = 13.7$ mm (using the second pulse echo) ellipticity reached a value of 0.8 and the rotation angle $\pi/2$ at their maxima; however, the shapes of the curves exhibited essential distortion. This property was the subject of further study [68, 190].

*Resonance line shape.* Measurements of the absorption $\Delta\Gamma^{\pm}(H)$ using different path lengths yielded the same values, even at high frequencies

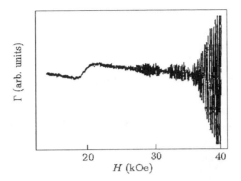

FIGURE 6.29. Magnetic field dependence of ultrasonic attenuation in cadmium. $f = 29\,\mathrm{MHz}$, $T = 2\,\mathrm{K}$, $l = 0.4\,\mathrm{cm}$. The variation of the curve at $H \approx 20\,\mathrm{kOe}$ is due to DPR. $\mathbf{H} \parallel \mathbf{k} \parallel [001]$, $\mathrm{RRR} = 10^5$. (After Fig. 3 in [28].)

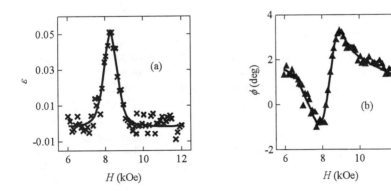

FIGURE 6.30. Magnetic field dependences of (a) ellipticity, and (b) the angle of rotation of the polarization in tungsten. $[100] \parallel \mathbf{H} \parallel \mathbf{k}$, $f = 54.3\,\mathrm{MHz}$, $T = 4.2\,\mathrm{K}$, $l = 4.57\,\mathrm{mm}$. (After Fig. 1 in [36].)

(up to 262 MHz). Since the method used was based on the assumption that polarization phenomena can be interpreted in terms of two circularly polarized modes [one (+) and one (−)], and $\Delta\Gamma^\pm(H)$ was independent of the path length, it became clear that the elastic and electromagnetic subsystems are weakly coupled in tungsten and equation (4.92) should be used to interpret the experimental data.

Model calculations [190] were carried out to see whether the line shapes of $\Gamma^\pm(H)$ and $s^\pm(H)$ can be explained by the existing theory. The starting point of all model calculations in metals is the FS, which is based on the expression for the energy of the conduction electrons as a function of quasi-momentum: $\mathcal{E} = \mathcal{E}(\mathbf{p})$.

The FS of tungsten was given in Fig. 6.14. It is rather complicated and consists of an electron jack, a hole octahedron, and small hole ellipsoids. It is very difficult to precisely describe such a surface analytically. Fortunately, the DPR under consideration here is determined by charge carriers having the largest resonance field value. (Keep in mind that it depends upon $\partial S/\partial p_z$.) Galkin et al. [31] proposed that it was caused by the conduction electrons situated near the limiting point on the jack, which he labeled $G$. He called the corresponding wave the $G$-doppleron (Fawcett et al. [191] labeled this point $S$). In fact, the problem of the location of the resonant charge carriers is still open. As will be seen, calculations based on the proposition that only the area near the point $G$ is responsible for DPR in tungsten do not give quantitative agreement with the experimental data. However, many important conclusions can be reached based on this proposition.

On the basis of this model, the contribution to the electroacoustic coefficients by the electrons near the limiting point $G$ should be described in detail, while contributions from the rest of the jack as well as the other sheets of the FS may be treated in the local approximation: The resonance field for the electrons near $G$ is so large that the other charge carriers will not produce any non-monotonic field dependences of $\alpha^\pm$, $\beta^\pm$, or $\sigma^\pm$. In addition, the absence of multiple resonances in tungsten indicates that the cyclotron orbits of the resonant electrons are essentially circular.

Thus a FS model with cylindrical symmetry about the $z$ direction was chosen. It is represented by two semispheroids having concavities near the limiting points. The possible existence of the concavities was pointed out in experiments performed in molybdenum [192], which has a similar FS. Figure 6.31(a) shows the FS model cross section in the half space $p_z > 0$.

The energy of the electrons as a function of quasi-momentum in the range of $p_1 \leq |p_z| \leq p_2$ is assumed to be given by

$$\mathcal{E}(\mathbf{p}) = (1/2m)\left\{ [p_z - p_c\,\mathrm{sgn}(p_z)]^2 - p_\perp^2 + p_\perp^4/2p_0^2 \right\},$$

$$(6.28)$$

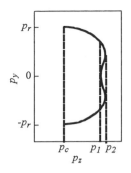 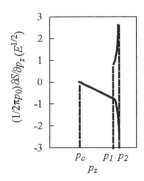

FIGURE 6.31. (a) Cross section of the tungsten Fermi surface model by the plane $p_x = 0$, and (b) $\partial S/\partial p_z$ in units of $E^{1/2}$ as a function of $p_z$. (After Fig. 1 in [190].)

where $p_\perp^2 = p_x^2 + p_y^2$. As usual, the FS is defined by the surface of constant energy $\mathcal{E}(\mathbf{p}) = \mathcal{E}_F$. In the range $p_c \leq |p_z| \leq p_1$ the model FS was assumed to be a sphere with radius $p_r = 0.27\hbar \times 10^8 \, \mathrm{g \, cm \, s^{-1}}$. The conditions for continuous differentiability of $S(p_z)$ at $p_z = \pm p_1$ give

$$\mathcal{E}(\mathbf{p}) = (1/2m)\left\{ [p_z - p_c \, \mathrm{sgn}(p_z)]^2 + p_\perp^2 - 2p_0^2 \right\} \qquad (6.29)$$

for the spherical part of the model. The parameter $m$ is defined from the condition $m_c(p_z) = m_{exp}$, where $m_c(p_z)$ is the cyclotron mass, $m_{exp} = 0.588 \, m_0$ is its experimentally determined value on the central cross section of the jack spheroid, and $m_0$ is the free electron mass. The value of $p_c$, which gives the position of the spheroid center with respect to the center of the Brillouin zone, and the values of $m_{exp}$ and $p_r$ were taken from [193]. The parameter $p_0$ is found from

$$p_0^2 = p_r^2 / (2 + E), \qquad (6.30)$$

where $E$ is a fitting parameter which determines the concavity of the model. In the calculations $E = 2.5$ was used.

This model was selected for the following reasons: The model FS is similar to that of tungsten near the limiting point $G$ and has the same linear dimensions. The derivative $\partial S/\partial p_z$, as a function of $p_z$ (see Fig. 6.31), has neither an extremum nor a discontinuity for finite values of $S$. The only anomaly of the derivative is the infinity at the point $p_z = p_2$. This is the result of the infinite magnitude of $m_c$ for the corresponding electrons. However, such electrons only contribute to $\alpha^\pm$, $\beta^\pm$, and $\sigma^\pm$ in the infinite limit of $H$. The existence of dopplerons is usually connected with the following features of $\partial S/\partial p_z$: (i) An extreme value or discontinuity for a finite interval of $S$, or (ii) a parabolic limiting point (constant value of $\partial S/\partial p_z$ in a finite interval of $p_z$). The FS of tungsten does not display these features,

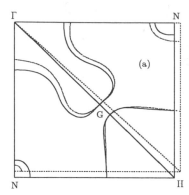
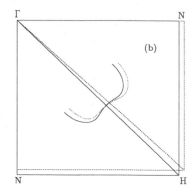

FIGURE 6.32. (a) Angular shear response of the FS of tungsten in the (100) plane perpendicular to the shear axis (after [191] p. 316). (b) The same response for the model under consideration. The broken curves correspond to deformations $\partial u_y/\partial z = \partial u_z/\partial y = 0.026$. (After [190].)

so it is of interest to discover whether a doppleron can exist under such conditions and to find out what new features this DPR possesses. In addition, the model makes possible an analytical description of $m_c$, $\mathbf{v}$, and, with some assumptions, even the deformation potential tensor $\boldsymbol{\lambda}$ introduced by Akhiezer [97].

In the calculations of coefficients that depend on the deformation potential (e.g., of $\alpha_{ijkl}$ and $\beta_{ijk}$), $\lambda_{mn}$ was postulated in the usual way (see [29, 181, 194]). However, in [190] a method of checking whether the choice of the deformation potential was successful was suggested. It was based on data for the FS of a metal that has undergone homogeneous deformations.

To do this one should write the expression for the electron energy in the laboratory system for zero electromagnetic fields, i.e., transform equation (4.64) into

$$\mathcal{E}\left(\mathbf{p}, \mathbf{u}\right) = \mathcal{E}\left(\mathbf{p}, 0\right) + \left[\lambda_{ij}\left(\mathbf{p}\right) + p_i v_j\right] \frac{\partial u_i}{\partial x_j} + \left(p_i - m_0 v_i\right) \frac{\partial u_i}{\partial t}, \tag{6.31}$$

then consider the static case $\partial \mathbf{u}/\partial t = 0$ and plot the surface of constant energy $\mathcal{E}\left(\mathbf{p}, \mathbf{u}\right) = \mathcal{E}_F$ with the chosen $\lambda_{ij}$. A comparison of this surface with the data mentioned before will reveal whether the choice has been correct.

In Ref. [190] the deformation parameter was taken to be given by

$$\lambda_{ij} = -\zeta v_i p_j. \tag{6.32}$$

The data obtained by Fawcett et al. [191] and presented in Fig. 6.32(a) show that there is good agreement when $\zeta = 3$. This value was used for the calculations of the electroacoustic coefficients $\alpha^{\pm}$ and $\beta^{\pm}$.

Since $|\partial S/\partial p_z|$ in the vicinity of the limiting point $G$ is greater than at any other point on the FS, the contributions of the rest of the electron jack and the other sheets to the electroacoustic coefficients are monotonic in the region of the DPR under consideration. Thus their contributions to $\alpha^{\pm}$ and $\beta^{\pm}$ were neglected, and $\sigma^{\pm}$ was obtained from the requirement that the static Hall conductivity vanish in a compensated metal to avoid the appearance of helicon–phonon resonances in the model calculations.

By analogy with equation (4.97), the nonlocal conductivity is

$$\sigma^{\pm} = \mp i \frac{4\pi c |e| p_0^3}{H h^3} \left\{ \mathrm{Re}\left[ F_\Sigma(q)\right] \pm i\mathrm{Im}\left[ F_\Sigma(q)\right] \right\}, \tag{6.33}$$

where

$$F_\Sigma(q) = F_\sigma(q) - \mathrm{Re} F_\sigma(0), \tag{6.34}$$

the nonlocality parameter is

$$q = \frac{k_0 c p_0}{|e| H}, \tag{6.35}$$

and the contribution of the resonant charge carriers is

$$\begin{aligned}
F_\sigma(q) = & \int_0^{x_1} dx \frac{(E + 2 - x^2)(1 - i\gamma)}{(1 - i\gamma)^2 - q^2 x^2} \\
& + \int_{x_1}^{x_2} dx \frac{\left(1 + \sqrt{1 - 2x^2 + 2E}\right)\left(1 - i\gamma/\sqrt{1 - 2x^2 + 2E}\right)}{\left(1 - i\gamma/\sqrt{1 - 2x^2 + 2E}\right)^2 - q^2 x^2/(1 - 2x^2 + 2E)} \\
& - \int_{x_1}^{x_2} dx \frac{\left(1 - \sqrt{1 - 2x^2 + 2E}\right)\left(1 + i\gamma/\sqrt{1 - 2x^2 + 2E}\right)}{\left(1 + i\gamma/\sqrt{1 - 2x^2 + 2E}\right)^2 - q^2 x^2/(1 - 2x^2 + 2E)},
\end{aligned} \tag{6.36}$$

where $x = (p_z - p_c)/p_0$, $x_1 = (p_1 - p_c)/p_0$, $x_2 = (p_2 - p_c)/p_0$, and $\gamma = mc/(H|e|\tau)$.

Figure 6.33 shows that $\mathrm{Re}[F_\Sigma(q)]$ has an extremum in the vicinity of $q = q_1$, while $\mathrm{Im}[F_\Sigma(q)]$ drops sharply as $q$ is reduced below $q_1$. This corresponds to the DSCR condition for electrons having $p_z = \pm p_1$, which should be called the effective ones in this case. An interesting feature of this FS model is the absence of a well-resolved absorption edge for ultrasonic waves (determined by $\mathrm{Im}[F_\Sigma(q)]$) analogous to the Kjeldaas edge for ultrasonic waves. This is a consequence of the infinity of $\partial S/\partial p_z$ that makes it possible to find electrons contributing to the collisionless cyclotron absorption for any magnetic field.

Using (4.72)–(4.74), (4.80), and (4.81) and integrating over the model FS, equations were derived for the other electroacoustic coefficients analogous to (6.33):

$$\alpha^{\pm} = \mp i \frac{4\pi c p_0^5}{|e| H h^3} \left(\frac{\varsigma}{2}\right)^2 \left[\mathrm{Re} F_\alpha(q) \pm i\mathrm{Im} F_\alpha(q)\right], \tag{6.37}$$

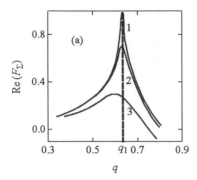
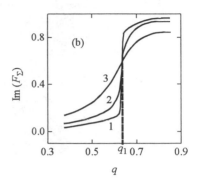

FIGURE 6.33. (a) Real and (b) imaginary parts of $F_\Sigma$ plotted as functions of $q$ for different values of relaxation time: (1) $\tau \to \infty$, (2) $\tau = 5 \times 10^{-10}$ s, and (3) $\tau = 10^{-10}$ s; $f = 100$ MHz. (After Fig. 3 in [190].)

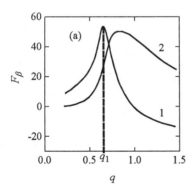
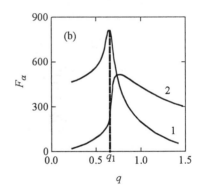

FIGURE 6.34. (1) Real and (2) imaginary components of $F_\alpha(q)$ and $F_\beta(q)$ as functions of $q$. $f = 100$ MHz, $\tau = 5 \times 10^{-10}$ s. (After Fig. 4 in [190].)

$$\beta^\pm = -i\frac{4\pi c p_0^4}{H h^3}\left(\frac{\varsigma}{2}\right)[\mathrm{Re}F_\beta(q) \pm i\mathrm{Im}F_\beta(q)]. \qquad (6.38)$$

The plots of $F_\alpha(q)$ and $F_\beta(q)$ presented in Fig. 6.34 are similar to those of $F_\Sigma(q)$ because the numerators of the integrals (4.72) consist of various smooth functions, whereas the denominator is common to all of the expressions and it is the denominator that provides the dominant contribution of the effective electrons to $\alpha^\pm$, $\beta^\pm$, and $\sigma^\pm$ when $\Omega\tau \gg 1$.

Inserting the expressions for the electroacoustic coefficients into equation (4.92) (whose validity was demonstrated experimentally for this case), the magnetic field dependences of the absorption and the phase velocity of circularly polarized elastic-like modes shown in Fig. 6.35 and Fig. 6.36, respectively, were obtained.

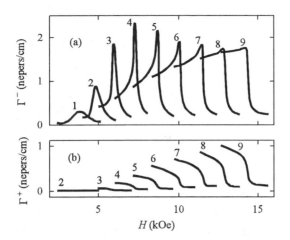

FIGURE 6.35. Calculated magnetic field dependence of the absorption of circularly polarized elastic-like modes, (a) $(-)$ polarization, (b) $(+)$ polarization. Curves $1 \leq n \leq 9$ are for $f = (n+1)50\,\text{MHz}$; $\tau = 5 \times 10^{-10}\,\text{s}$. (After Fig. 5 in [190].)

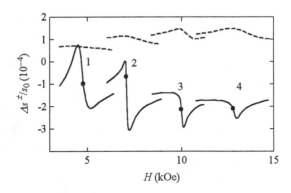

FIGURE 6.36. Calculated magnetic field dependence of the phase velocity of circularly polarized elastic-like modes, where $\Delta s^{\pm} = s^{\pm}(H) - s_0$. Curves 1–4 are for $f = 150$, 250, 350, and 450 MHz, respectively. Solid curves, $(-)$ polarization; broken curves, $(+)$ polarization; $\tau = 5 \times 10^{-10}\,\text{s}$. The filled circles indicate the zero level of $\left[ s^{-}(H) - s^{-}(H_R^{-}) \right] / s_0$. (After Fig. 6 in [190].)

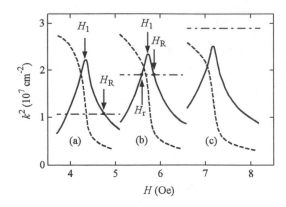

FIGURE 6.37. Plots of $k_0^2$ (horizontal broken lines), $\mathrm{Re}(k_E^-)^2$ (solid lines), and $-\mathrm{Im}(k_E^-)^2$ (dashed lines) as functions of the magnetic field at fixed frequencies: (a) $f = 150$, (b) 200, and (c) 250 MHz. The arrows indicate the magnetic field values for the fundamental DPR ($H_R^-$), DSCR ($H_1$), and the additional DPR ($H_r^-$); $\tau = 5 \times 10^{-10}$ s, $\delta = \mathrm{Re}\left[k_0^2 - (k_E^-)^2\right]$. (After Fig. 7 in [190].)

As can be seen, DPR occurs for the $(-)$ polarization and results in the appearance of an absorption peak and the well-known shape of the phase velocity dispersion. Beginning at $f = 50$ MHz the peak grows with increasing frequency, becoming slightly asymmetrical (the high-field wing is broadened). A further frequency increase leads to a broadening of the low-field wing of the absorption peak. When $f > f_L$ ($f_L = 225$ MHz), the DPR peculiarities in $\Gamma^-$ diminish and further growth of the absorption peak (for $f > 500$ MHz) is caused by DSCR in both the polarizations.

Let us discuss the reasons for this kind of variation in the line shape. The line shape of the absorption peak is determined by the magnetic field dependences of the numerator and denominator of the field term in the right-hand side of equation (4.92). In the long-wavelength range ($f < 200$ MHz) the value of the DPR field $H_R^-$ differs significantly from $H_1$. Note that $H_R^-$ corresponds to the absorption maximum and is determined by the zero in the magnitude of $\mathrm{Re}[k_0^2 - (k_E^-)^2]$, while $H_1$ is the DSCR field for the effective electrons, i.e., those with $p_z = p_1$. Therefore the direct influence of DSCR on the DPR line shape is small and the numerator in equation (4.72) does not exhibit any appreciable variation in the vicinity of DPR. As for the denominator, $\mathrm{Re}[k_0^2 - (k_E^-)^2]$ changes at different rates as the magnetic field decreases or increases from $H_R^-$ (see Fig. 6.37, $f = 150$ MHz). Specifically, for $H < H_R^-$ the more rapid increase of $|\mathrm{Re}[k_0^2 - (k_E^-)^2]|$ results in the narrowing of the low-field wing of the absorption peak in comparison with the high-field one. The narrowing of the low-field wing in the long-wavelength region is a feature of DPR determined by the doppleron having a phase velocity $\mathbf{v}_E$ parallel to, but oppositely directed from, the group velocity $g_E$. When the FS has a sufficiently complicated shape, the presence of a minimum in $|\partial S/\partial p_z|$ is possible. This results in the existence of a doppleron with $v_E$ in the same direction as $g_E$ (see [122]). In this case the magnetic field dependence of $\mathrm{Re}(k_E^2)$ is inverted with respect to the line $H = H_1$ providing a narrowing of the high-field wing of the absorption peak. Thus, the line shape of the absorption peak in the long-wavelength region of DPR indicates the type of doppleron observed.

In the middle wavelength range ($200 \leq f \leq 225\,\mathrm{MHz}$) the widening of the low-field wing is due to two circumstances: (1) The DPR is located near the DSCR causing ultrasonic absorption at $H < H_1$ ($q > q_1$), and (2) in this frequency range there are two closely spaced points where the real component of the denominator in the field term vanishes (see Fig. 6.37, $f = 200\,\mathrm{MHz}$). Formally, this corresponds to two DPRs: A principal resonance at $H = H_R^-$ and an additional one at $H = H_r^-$. In the model being discussed the zero of $\mathrm{Re}[k_0^2 - (k_E^-)^2]$ at the lower field occurs when $|\mathrm{Im}[k_0^2 - (k_E^-)^2]|$ is sufficiently large and does not result in a separate absorption peak, though it broadens the low-field wing of the basic DPR.

Since $k_0^2$ increases quadratically with $f$, and $\mathrm{Re}[(k_E^-)^2]$ is bounded, it is possible to reach a frequency $f_L$ where the line $k_0^2$ will touch the curve $\mathrm{Re}[(k_E^-)^2]$ at its maximum. In the short-wavelength region ($f > f_L$), $\mathrm{Re}[k_0^2 - (k_E^-)^2]$ never vanishes as the line $k_0^2$ lies above the curve $\mathrm{Re}[(k_E^-)^2]$ (see Fig. 6.37, $f = 250\,\mathrm{MHz}$). So $f_L$ is the limiting value of the frequency where DPR may be observed. The appearance of a limiting frequency for the observation of DPR is due to a finite value of the nonlocal conductivity. In the model under discussion this feature of $\sigma^{\pm}$ originates from the existence of cyclotron absorption for any magnetic field, even in the collisionless limit.

The line shape of the phase velocity dispersion can be seen in Fig. 6.36, where the anomaly corresponding to DPR appears on the smooth curve caused by DSCR. One can see from equation (4.92) that the field term contribution to the phase velocity dispersion of the $(-)$-polarized mode vanishes at $H = H_R^-$. Hence, it is reasonable to calculate the phase velocity variation relative to its value at $H = H_R^-$: $\delta s^- = s^-(H) - s(H_R^-)$.

The model calculations presented here have been compared with the experimental data (see Figs. 6.38 and 6.39) obtained with the use of specimens that had RRR $> 1.5 \times 10^5$. The main features of the absorption and phase velocity dispersion of circularly polarized ultrasonic-like modes experimentally observed in tungsten under DPR conditions agreed qualitatively with calculations based on the suggested model. Quantitative agreement (for example, in magnitudes of the resonant fields, the value of the limiting frequency and those separating the short-, middle-, and long-wavelength regions) requires a more accurate FS and, in particular, derivative of its cross section.

Nevertheless, it was shown that the absorption peak has an asymmetric shape and in the low-frequency range of DPR observation the low-field wing is narrower than the high-field wing and vice versa for higher frequen-

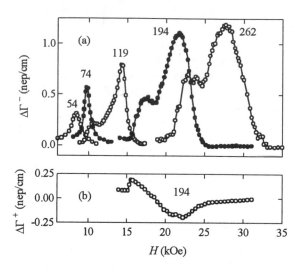

FIGURE 6.38. Magnetic field dependence of the absorption of circularly polarized elastic-like modes in tungsten with $[100] \parallel \mathbf{H} \parallel \mathbf{k}$, RRR $> 1.5 \times 10^5$, $T = 4.2\,\mathrm{K}$. $\Delta\Gamma^\pm = \Gamma^\pm(H) - \Gamma^\pm(2H_R)$, where $H_R$ is the resonance field. The frequencies in megahertz are given by numbers near the curves. (a) $(-)$ polarization, (b) $(+)$ polarization. (After Fig. 8 in [190])

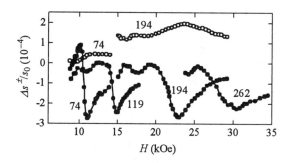

FIGURE 6.39. Magnetic field dependence of the phase velocity of circularly polarized elastic-like modes in tungsten, where $\Delta s^\pm = s^\pm(H) - s_0$. Open circles, $(+)$ polarization; closed circles, $(-)$ polarization. Experimental details same as in Fig. 6.38. (After Fig. 9 in [190].)

cies. Owing to the restrictions of their experimental technique, the authors failed to reach the frequency $f_L$ where the DPR anomalies stop increasing; however, the tendency of $\Delta\Gamma^-$ peak growth to diminish for $f > 119\,\text{MHz}$ was observed.

The existence of the limiting frequency $f_L$ for the model considered has a wider sphere of application than merely the explanation of some DPR features typical for this specific model or for DPR in tungsten. Since the limiting frequency originates from the finite value of nonlocal conductivity, and the latter is always finite in real metals at least due to finite relaxation time, there will always be a highest frequency at which the doppleron and phonon dispersion curves cross and therefore at which DPR occurs. The existence of such a crossing is the starting point for the method of reconstructing the nonlocal Hall conductivity introduced by Butenko et al. [195].

This problem as well as the occurrence of a small maximum in the experimental curves of ultrasonic absorption for (+) polarization (labeled $H_r^+$ in Fig. 6.38) were studied by Gudkov and Zhevstovskikh in [196] and [68].

*Pseudo-doppleron–phonon resonance.* From careful measurements of the positions of the absorption peaks at a number of fixed frequencies, it was found that $H_r^-$ and $H_r^+$ are different at a given frequency and deviate from a linear dependence on the magnetic field as $f$ is varied. These circumstances rule out an assignment of these features to peaks in the deformation absorption.

To clarify the situation, a rather simple model of the FS was considered in Ref. [195]. The electron sheet had the shape of a "truncated sphere," while the hole sheet was a cylinder of the same volume. The radius of the sphere was chosen to be $p_r = 0.27h \times 10^8\,\text{g\,cm\,s}^{-1}$ and that of the limiting cross-section $p_0 = 0.179h \times 10^8\,\text{g\,cm\,s}^{-1}$. The magnetic field dependences of the components of the nonlocal conductivity $\sigma_{21}(k, H)$ and $\sigma_{11}(k, H)$ are given for such a model in Fig. 6.40.

The position of the DPR is determined by the vanishing of the real part of the denominator in the field term, $[k_0^2 - (k_E^-)^2]$. This condition, however, can be satisfied for both polarizations. Let us rewrite it as

$$\left(k_0^2 c^2 / 4\pi\omega\right) - (\text{sgn}\,P)\sigma_{21}(k_0, H) = 0, \tag{6.39}$$

where $\text{sgn}\,P = \pm 1$ for the $(\pm)$ polarizations. The graphical solution of the equation (6.39) is the intersection of the curve $\sigma_{21}(k, H)$ with the lines $\pm k_0^2 c^2 / 4\pi\omega$. The intersection which occurs for

$$|\sigma_{21}| \gg |\sigma_{11}| \tag{6.40}$$

relates to the resonance caused by the interaction of ultrasound with a weakly damped electromagnetic wave (doppleron, in our case). In addition, the intersections of $\sigma_{21}(k, H)$ with the lines $\pm k_0^2 c^2 / 4\pi\omega$ take place in the region of cyclotron absorption (i.e., for $H < H_1$, where $H_1$ as above corresponds to DSCR of charge carriers having a maximum in $|m_c v_z|$).

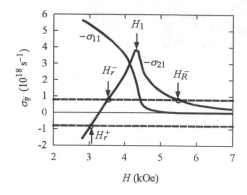

FIGURE 6.40. Magnetic field dependences of components of the nonlocal conductivity for a FS model with the shape of a "truncated sphere." The horizontal dashed lines are drawn for $\pm k_0^2 c^2 / 4\pi\omega$, where $k_0 = 2\pi f / s_0$. Here $f = 150\,\text{MHz}$, $s_0 = 2.88 \times 10^5\,\text{cm/s}$, and $\tau = 5 \times 10^{-10}\,\text{s}$. (After Fig. 2 in [68].)

The vanishing of the real part of the field term denominator will result in an increase in the electromagnetic absorption of ultrasound, and the structure of this increase can have the form of a resonant peak under certain conditions (to be analyzed later). Figure 6.41 shows the behavior of the ultrasonic wave number and the positions of the resonant field in the coordinate system usually used for presenting doppleron dispersion curves.

It can be seen that the dependences given in Fig. 6.41 differ from the linear ones typical for peaks of deformation absorption. In the long-wavelength region the curves of electromagnetic absorption for $H < H_1$ (i.e., Curves 2 and 3) asymptotically approach the line which passes through the origin, while in the short-wavelength region they bend away from this line. Moreover, Curve 2, which corresponds to the peak in $(-)$ polarization, meets Curve 1 (relating to DPR) at a point located on the line $k_0(v_z)_{\max} = \Omega$ (DSCR of the electrons on the limiting cross-section of the FS).

In Fig. 6.42 experimental data are shown for the magnetic field of the principal peak and two additional ones measured at a number of frequencies and given in the same coordinate system as used in Fig. 6.41. These curves exhibit similarity with the calculated ones shown in Fig. 6.41. The quantitative differences in the values of magnetic field are due to the fact that the magnitude of $|m_c v_z|$ is too small in this model.

Thus the analysis of the calculated and experimental curves describing the positions of the peaks in $(H, f)$ coordinates shows that the peaks of $\Gamma^\pm$ that occur in addition to DPR can be assigned to peaks of electromagnetic absorption in the region of DSCR. In appearance and formal description these peaks are similar to DPR; however, they are observed for a relation between nondissipative (Hall) and dissipative components of the conduc-

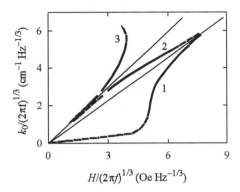

FIGURE 6.41. Relation between $k_0$ and magnetic field corresponding to equation (6.39) for $\sigma_{21}$ as given in Fig. 6.40. Curve 1 is for $(-)$ polarization, $H > H_1$; Curve 2 for $(-)$ polarization, $H < H_1$; Curve 3 for $(+)$ polarization, $H < H_1$. Solid portions of the curves correspond to the frequency interval used in the experiment. (After Fig. 3 in [68].)

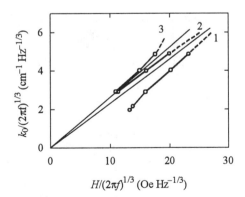

FIGURE 6.42. Relation between $k_0$ and resonant magnetic field, measurements in tungsten indicated by circles. Curve 1 is for $(-)$ polarization, principal resonance at $H_R^-$, Curve 2 for $(-)$ polarization, additional peak at $H_r^-$, and Curve 3 for $(+)$ polarization, additional peak at $H_r^+$. (After Fig. 4 in [68].)

tivity tensor,

$$|\sigma_{21}| < |\sigma_{11}|, \tag{6.41}$$

which is unusual for dopplerons.

Since the existence of DPR as well as the additional peaks have the same cause, namely, a sharp extremum in Hall conductivity due to DSCR of a certain group of charge carriers, it is possible to use the term "satellites" for the additional peaks. Actually, the satellites are pseudo-DPR, because the magnitude of $k = k_0$ that causes the real part of the field term denominator to vanish at $H = H_r^{\pm}$ is not a solution of the dispersion equation that relates to a weakly damped electromagnetic wave, contrary to the condition for a real doppleron.

*Nonlocal conductivity reconstruction.* The discovery of pseudo-DPR made it possible to expand the method of nonlocal Hall conductivity reconstruction [195] to the region of magnetic field below the Kjeldaas edge. Furthermore, in [68] it was shown how to estimate the dissipative component of conductivity above the Kjeldaas edge by measuring the width of the high-field wing of the ultrasonic absorption peak related to DPR.

As in equation (4.97), nonlocal conductivity was presented as the static Hall conductivity $\sigma_0 = N_e|e|c/H$, where $N_e = N_h$ are the concentrations of electrons and holes in tungsten, multiplied by the normalized nonlocal conductivity expressed by the complex function $F(q,\gamma) = F'(q,\gamma) + i\,F''(q,\gamma)$:

$$\sigma^{\pm}(k_0, H) = i\,\sigma_0\left[\pm F'(q,\gamma) + i\,F''(q,\gamma)\right], \tag{6.42}$$

where $\gamma = 1/\Omega_e\tau_e$, the subscript $e$ indicates that the value relates to the resonant electrons, and $q$ is the parameter of nonlocality, defined by

$$q = \frac{k_0 c}{2\pi|e|H}\left(\frac{\partial S}{\partial p_z}\right)_e. \tag{6.43}$$

Butenko et al. [195] neglected the dependence of $F$ on $\gamma$ while reconstructing the nonlocal conductivity, assuming that the collisionless limit had been reached already at the temperature of the experiment (4.2 K). In [68] it was found that lowering the temperature to 1.8 K resulted in a noticeable rise in the resonant peaks of ultrasonic absorption. This was an indication that the collisionless limit had not been reached so that it was incorrect to neglect the collisions when determining the nonlocal conductivity. The result of reconstructing $F'$, to be denoted by subscript $s$, should be a curve on the surface $F' = F'(q,\gamma)$. To obtain $F'_s$ one must transform equation (6.39), taking into account the definition (6.42):

$$F'_s = (\text{sgn}\,P)\frac{cfH_R^-}{2s_0^2 N_e|e|}. \tag{6.44}$$

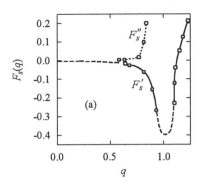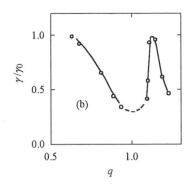

FIGURE 6.43. (a) Projections of $F_s'(q,\gamma)$ and $F_s''(q,\gamma)$ onto the $\gamma = \text{const}$ plane. (b) Projection of $F_s'(q,\gamma)$ onto the $F' = \text{const}$ plane. Circles represent measurements in tungsten. (After Fig. 5 in [68].)

The value of the nonlocality parameter at which $F_s'$ is computed is given by

$$q = \frac{\omega\theta^{-1}}{s_0 H_R^-}, \qquad (6.45)$$

where $\theta^{-1}$ is the slope of the asymptote $k_0(v_z)_e = \Omega_e$ approached by the DPR position in the $(H/\omega^{1/3}, k_0/\omega^{1/3})$ coordinate system:

$$\theta^{-1} = \frac{c}{2\pi|e|}\left(\frac{\partial S}{\partial p_z}\right)_e = \frac{c}{2\pi|e|}(m_c v_z)_e. \qquad (6.46)$$

Since the method of reconstructing the nonlocal conductivity is based on equation (6.39), which is valid for the magnetic fields of the satellites as well, it is also possible to apply it in the DSCR region $(H < H_K)$. In this case equations (6.44) and (6.45) are still valid after replacing $H_R^-$ by $H_r^\pm$.

Figure 6.43 shows the projections of the curve $F_s' = F_s'(q,\gamma)$ on the planes $\gamma = \text{const}$ and $F' = \text{const}$ obtained for $\theta^{-1} = 4.42\,\text{Oe cm}$. The broken curve for $q < 0.6$ was plotted by taking into account the vanishing of the Hall conductivity in the quasi-static limit,

$$\lim_{q,\gamma\to 0} F'(q,\gamma) = 0. \qquad (6.47)$$

In addition to reconstructing $F_s'(q,\gamma)$, Gudkov and Zhevstovskikh [68] introduced a method of evaluating $F_s''(q,\gamma)$. It was applied to tungsten in which all of the resonant features of such electroacoustic coefficients as $\alpha^\pm$ and $\beta^\pm$ occur at magnetic fields less than that for DPR (i.e., at $H < H_R^-$). In addition, Vlasov et al. [181] showed that the contribution to $\Gamma^\pm(H)$ of the deformation term $(\text{Re}\,\alpha^\pm)$ for fixed $k_0$ did not depend on

$H$. Thus, if we measure the resonant curve relative to its high-field limit, $\Delta\Gamma^-_{res} = \Gamma^-(H) - \Gamma^-(\infty)$, where $\Gamma^-(\infty)$ is the value measured at $H \gg H^-_R$, we can write the expression for $\Delta\Gamma^-_{res}$ at fields above the Kjeldaas edge as

$$\Delta\Gamma^-_{res} = \frac{k_0^3}{2\rho\omega}\sigma_{11}\left(\text{Im}\beta^- - \frac{cHk_0}{4\pi\omega}\right)^2\left[\sigma_{11}^2 + \left(\frac{k_0^2c^2}{4\pi\omega} + \sigma_{21}\right)^2\right]^{-1}. \quad (6.48)$$

If we denote the width of the high-field wing of the resonant curve by $\Delta H$, evaluate $\Delta\Gamma^-_{res}$ at $H_\Delta = H^-_R + \Delta H$, divide it by $\Delta\Gamma^-_{res}(H^-_R)$, and move the terms containing $\sigma_{11}$ to the left-hand side, we get

$$\left[2B^2\sigma_{11}(k_0, H^-_R) - \sigma_{11}(k_0, H_\Delta)\right]\sigma_{11}(k_0, H_\Delta)$$
$$= \left[\sigma_{21}(k_0, H_\Delta) - \sigma_{21}(k_0, H^-_R)\right]^2, \quad (6.49)$$

where

$$B^2 = \frac{\left[\text{Im}\beta^-(k_0, H_\Delta) - \frac{cH_\Delta k_0}{4\pi\omega}\right]^2}{\left[\text{Im}\beta^-(k_0, H^-_R) - \frac{cH^-_R k_0}{4\pi\omega}\right]^2}. \quad (6.50)$$

It can be seen in equation (6.48) that the decrease in $\Delta\Gamma^-_{res}$ for $H > H^-_R$ is due to the decrease in the numerator of the field term and the increase in the difference $(k_0^2c^2/4\pi\omega) - |\sigma_{21}|$. If the last factor is dominant and the dependence of $\sigma_{11}$ on $H$ near $H^-_R$ is negligible (as in Fig. 6.40), for a narrow resonance (in tungsten, $\Delta H/H^-_R \leq 5\%$) it is possible to assume

$$B^2 \approx 1, \quad \sigma_{11}(k_0, H_\Delta) \approx \sigma_{11}(k_0, H^-_R), \quad (6.51)$$

and transform equation (6.49) into

$$\left[\sigma_{11}(k_0, H_\Delta)\right]^2 = \left[\sigma_{21}(k_0, H_\Delta) - \sigma_{21}(k_0, H^-_R)\right]^2. \quad (6.52)$$

In determining $\sigma_{11}(k_0, H_\Delta)$ we neglect the dependence of $F'$ on $\gamma$ in the interval $H^-_R \leq H \leq H_\Delta$. Actually, this means that we use the projection of $F'_s$ on the plane $\gamma = \text{const}$. As for the real part of $F(q, \gamma)$, the graphical representation of its imaginary part is a surface $F''(q, \gamma)$ in a three-dimensional space, and it is possible to reconstruct only a line $F'' = F''_s(q, \gamma)$ on this surface from an ultrasonic experiment. Note that the projections of the experimental curves $F' = F'_s(q, \gamma)$ and $F'' = F''_s(q, \gamma)$ on the $q - \gamma$ plane differ slightly since the first one is determined by the position of the DPR ($H^-_R$), while the second is determined by the field corresponding to the high-field wing ($H_\Delta$), i.e., $F' = F'_s(q_R, \gamma_R)$, but $F'' = F''_s(q_\Delta, \gamma_\Delta)$, where the subscripts on $q$ and $\gamma$ coincide with the subscripts on $H$. Using the definition of normalized nonlocal conductivity (6.42) we have

$$F''_s(q_\Delta, \gamma_\Delta) \approx F'_s(q_\Delta) - \left(H_\Delta/H^-_R\right)F'_s(q_R). \quad (6.53)$$

The projection of the curve $F'' = F_s''(q, \gamma)$ on the plane $\gamma = \text{const}$ for $q < 1$ is given in Fig. 6.43. Note that, since we have used the approximation (6.51), the data obtained should be regarded as estimates. For fields below the Kjeldaas edge the line shape of the absorption curve depends on more factors and therefore it is impossible to get even estimates of $F_s''(q, \gamma)$ in this region.

Let us compare two methods of reconstructing the nonlocal conductivity: using the radio-frequency size effect (also described in [195]) and the one presented here. In the first case $F'(q)$ is determined by the dispersion curves of dopplerons. The dependence of the doppleron wave number $\text{Re}(k_E)$ on $H$ at $\omega = \text{const}$ is determined by the oscillations of the surface impedance due to doppleron excitation. The condition for the oscillations' existence is a low level of doppleron absorption,

$$|\text{Im}(k_E)| \ll |\text{Re}(k_E)|, \tag{6.54}$$

or, in terms of conductivity,

$$|\sigma_{11}/\sigma_{21}| = |F''/F'| \ll 1. \tag{6.55}$$

At first glance, it would appear that the method based on DPR investigation should require the same conditions, since appreciable absorption of the doppleron leads to the broadening or even the disappearance of the resonant peak in ultrasonic absorption. However, this is not quite correct.

To analyze the conditions required for the existence of a rather narrow peak in ultrasonic absorption (i.e., $\Delta H/H_R^- \ll 1$), let us rewrite equation (6.52) with the assumption that $\sigma_{21}(k_0, H_\Delta)$ depends linearly on $H$ at $H = H_R^-$:

$$[\sigma_{11}(k_0, H_\Delta)]^2 \approx \left[ \Delta H \left( \frac{\partial \sigma_{21}}{\partial H} \right)_{H=H_R^-} \right]^2. \tag{6.56}$$

In the expression for the derivative

$$\frac{\partial \sigma_{21}}{\partial H} = -\frac{\sigma_0}{H} \left( F' + q \frac{\partial F'}{\partial q} + \gamma \frac{\partial F'}{\partial \gamma} \right) \tag{6.57}$$

it is possible to neglect the last term when $|\gamma| \ll 1$. As a result,

$$\frac{\Delta H}{H_R^-} \approx \left| \frac{\sigma_{11}(k_0, H_\Delta)}{H_R^- (\partial \sigma_{21}/\partial H)_{H=H_R^-}} \right|$$

$$\approx \left| \frac{(H_R^-/H_\Delta) F''(q_\Delta, \gamma_\Delta)}{F'(q_R, \gamma_R) + q_R(\partial F'/\partial q)_{q=q_R, \gamma=\gamma_R}} \right|. \tag{6.58}$$

This equation shows that the width of the absorption resonance is proportional to the dissipative component of conductivity $F''$, as one would expect. Less expected is the presence of its derivative with respect to $q$ in the denominator on the right-hand side of equation (6.58). The derivative can provide a rather narrow width of the resonance even when $F''$ is of the same order as, or exceeds, $F'$. This is because this denominator shows how fast the curve $\sigma_{21}(k_0, H)$ and the line $k_0^2 c^2/4\pi\omega$ disperse near the crossover.

Thus it becomes clear why in tungsten at $q > 0.8$, in spite of the fact that $F'$ and $F''$ are of the same order, a well-resolved peak in the ultrasonic absorption is observed: The variation of the Hall conductivity due to nonlocal effects (DSCR) is very large.

The same qualitative considerations can be applied to satellites (pseudo-DPR) as well. At $H = H_r^{\pm}$ the magnitude of the derivative $\partial F'/\partial q$ is so large that it leads to ultrasonic absorption peaks when the inequality in equation (6.55) is reversed. It is interesting to note that the derivative $\partial F'/\partial q$ in equation (6.58) can not only sharpen the resonance, but can broaden it as well. The latter occurs when $F' \approx q_R \partial F'/\partial q$ and the graph of $\sigma_{21}(k_0, H)$ crosses the line $k_0^2 c^2/4\pi\omega$ at a very small angle.

From these considerations, one can conclude that the interaction of the electromagnetic and the elastic fields in a metal in the weakly coupled regime manifests itself as resonant peaks in the absorption of circularly polarized elastic-like modes. The narrow peaks can be observed when

$$\left| \frac{F''}{F' + q(\partial F'/\partial q)} \right| \ll 1. \tag{6.59}$$

In the case where the relation (6.55) is valid, such a peak should be regarded as a DPR, otherwise as a pseudo-DPR: a result of the interaction of ultrasound with the circularly polarized electromagnetic modes which are highly damped.

While discussing ellipticity and rotation of the polarization of ultrasound and the data on the dissipative and Hall components of conductivity in a metal [or, as in equation (6.54], relations between the real and imaginary parts of a doppleron's wave number) in general and in tungsten in particular, it is pertinent to mention a paper by Golik et al. [189]. These authors used a combined acousto-electromagnetic technique (see Fig. 6.44) and observed rotation of the polarization as shown in Fig. 6.45. They found the ellipticity of the ultrasound to be $\varepsilon \approx 0.1$ and the damping of the doppleron to be $(\Gamma_D)^{-1} = 1.3/k_0$ at $f \approx 155\,\text{MHz}$ and $H = 20\,\text{kOe}$. Their data on doppleron damping is in good agreement with those obtained for the components of the nonlocal conductivity shown in Fig. 6.43.

Let us note that the radio-frequency size effect, in addition to providing the Hall component of conductivity, enables one to perform DPR investigations as well. Medvedev et al. [29] showed that the surface impedance $Z = R + iX$ of a resonant circuit with a cadmium plate in the coil, obtained as a function of $H$, has a reduction in oscillation amplitude near a DPR (see Fig. 6.46) Thus, the position of the reduction at a fixed frequency indicates $H_R$ and enables one to determine (with known sound velocity) $q_R$

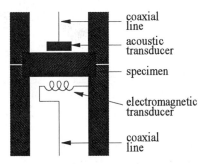

FIGURE 6.44. Schematic representation of an experiment employing an acousto-electromagnetic technique: Electromagnetic excitation of an ultrasonic mode and detection by a piezoelectric transducer. (After Fig. 1 in [188])

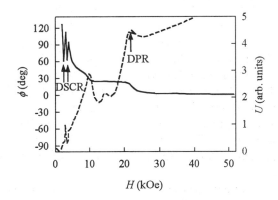

FIGURE 6.45. Rotation of the polarization of ultrasound (solid curve) and signal amplitude (dashed curve) as functions of magnetic field, obtained with the use of the acousto-electromagnetic technique in tungsten. [001] $\parallel$ **H** $\parallel$ **k**, $f \approx 155\,\text{MHz}$, $l = 2\,\text{mm}$. (After Fig. 6 in [189].)

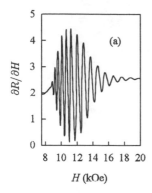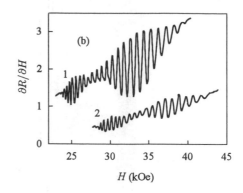

FIGURE 6.46. $\partial R/\partial H$ in arbitrary units as a function of magnetic field for a cadmium plate having $L = 0.57\,\mathrm{mm}$ and $[0001] \parallel \mathbf{H} \parallel \mathbf{k}_D$. $T = 1.5\,\mathrm{K}$, RRR$\approx 3 \times 10^4$. (a) Typical oscillations of the surface impedance of the resonant circuit due to doppleron excitation in the specimen, $f = 3\,\mathrm{MHz}$. (b) Reduced oscillation amplitude near each DPR: Curve 1, $f = 50\,\mathrm{MHz}$; Curve 2, $f = 73\,\mathrm{MHz}$. (After Fig. 5 in [29].)

and $F'$. Actually, to obtain $k_E$ (or $F'$) even in the region away from DPR, it is enough to know the dimension of the plate in the shortest direction and to have data such as given in Fig. 6.46(a). However, in view of the fact that impedance oscillations can only arise under the conditions given in equations (6.54) and (6.55), the ultrasonic method has a definite advantage: It enables one to reconstruct the nonlocal conductivity with the use of a pseudo-DPR [which requires the weaker condition (6.59)] as well.

*Is it actually necessary to take into account the ellipticity and rotation of the polarization while investigating DSCR or DPR by the ultrasonic technique?* Or is it possible to use a more traditional and simple method based on measuring the amplitude of the signal on the receiving transducer?

It is impossible to give a unique answer for all cases since it depends upon the particular situation. To take into account MPP means that one must process the measured data with the help of one of the methods described in Chapter 2.

However, as has been mentioned already, the first experiments were performed without accounting for MPP since at that time there was no idea that polarization phenomena would appear under DPR. Consequently, the experimental curves displayed peculiar forms, far from that of a typical resonant peak, as can be seen in Fig. 6.29.

Nevertheless, the result of such investigations was the reconstructed dispersion curve for the doppleron in cadmium given in Fig. 6.28. Model calculations demonstrated that the ratio of the radii which characterizes the electron lens is $r/R \approx 0.11$.

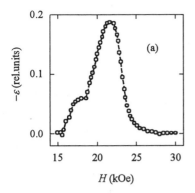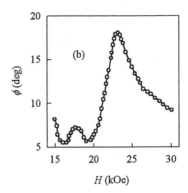

FIGURE 6.47. Magnetic field dependences of ellipticity and rotation of polarization obtained in tungsten at $f = 194\,\text{MHz}$ and $l = 0.29\,\text{cm}$. (After Fig. 6 in [68].)

Let us use an example to illustrate what can be obtained by accounting for and by neglecting MPP under DPR in tungsten where $\varepsilon$ and $\phi$ have the values presented in Fig. 6.47. The actual attenuation (in decibels) of a circularly polarized mode for which DPR is observed is related to the absorption $\Gamma^-$ and the path $l$ by the well-known expression

$$\Delta N(H) = N(H) - N(0) = 20 \log(e^{\Delta \Gamma^- l}), \tag{6.60}$$

whereas the apparent attenuation, measured by the traditional technique with the two transducers (receiving and generating) attached in parallel, is

$$\Delta N_m(H) = -20 \log\left[\left(\sqrt{A_1^2 + A_2^2}\right)/2\right], \tag{6.61}$$

where

$$A_1 = \exp\left[-\text{Im}(\Delta k^+)l\right] \cos\left[\text{Re}(\Delta k^+)l\right] \\ + \exp\left[-\text{Im}(\Delta k^-)l\right] \cos\left[\text{Re}(\Delta k^-)l\right],$$

$$A_2 = \exp\left[-\text{Im}(\Delta k^+)l\right] \sin\left[\text{Re}(\Delta k^+)l\right] \\ + \exp\left[-\text{Im}(\Delta k^-)l\right] \sin\left[\text{Re}(\Delta k^-)l\right].$$

It can be seen in Fig. 6.48 that, when the path length $l$ increases, the difference between $N_m(H)$ and $N^-(H)$ increases, not only in the values obtained at the same $H$, but in the shapes of the curves and in the position of the DPR peak. Thus, to determine the position of DPR and the width of the resonance in tungsten, it is necessary to use specimens that are as thin

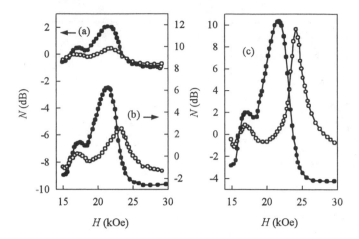

FIGURE 6.48. Magnetic field dependences of measured attenuation $N_m$ (open circles) and actual attenuation $N^-$ of the $(-)$ polarized mode (filled circles) in tungsten for different ultrasonic path lengths $l = (2n-1)L$, where $L = 0.29$ cm is the thickness of the specimen in the direction of propagation and $f = 194$ MHz. (a) $n = 1$; (b) $n = 2$; and (c) $n = 3$. (After Fig. 7 in [68].)

as possible, otherwise one would have to process the data to get curves for circularly polarized modes. As for satellites, they cannot be resolved without measuring the attenuation of circularly polarized waves.

*Multiple resonances.* Since dopplerons are caused by DSCR of an effective group of charge carriers and, for the case when the cyclotron orbit of these carriers is not circular, DSCR exhibits multiple harmonics, it was natural to expect multiple dopplerons and DPRs as well. Multiple dopplerons were discovered in aluminum by Skobov et al. [120] and multiple DPRs in indium by Vlasov and Gudkov [37]. The FSs of these metals are similar, so many important conclusions of the work in aluminum were used to interpret the data obtained for indium.

Skobov et al. [120] found two series of multiple dopplerons that were due to DSCR of two groups of effective charge carriers (holes in fact). One is located at a maximum of $|\partial S/\partial p_z|$ whereas the second is at a minimum (see Fig. 6.49). The FS model with cross-sectional area and derivative with respect to electron wave number given in Fig. 6.49 was used to interpret the experimental data. In addition, in order to discuss multiple harmonics of DSCR, the shape of the cross section of the model was chosen as an astroid:

$$k_x^{2/3} + k_y^{2/3} = [8S(k_z)/3\pi]^{1/3} . \qquad (6.62)$$

Of course, this does not accurately describe the real metal, but the cross section has fourth-order rotational symmetry with respect to the $z$ direction (and therefore the numbers of DSCR harmonics typical for a cubic axis

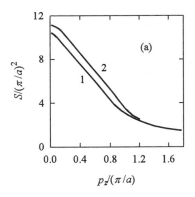
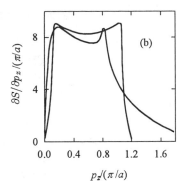

FIGURE 6.49. (a) $S(p_z)$ and (b) $|\partial S/\partial p_z|$ for the aluminum FS as functions of the electron wavevector $p_z$. Curves 1 and 3 are from calculations by Larsen and Greisen [197] while Curves 2 and 4 are from a model used by Skobov et al. [120], $a = 4.05\,\text{Å}$. (After Fig. 1 in [120].)

parallel to $\mathbf{H}$) and, in addition, is similar to the cross section of the FS of aluminum near the minimum of $|\partial S/\partial k_z|$.

The nonlocal conductivity for such a model can be written as a sum of multiple harmonics:

$$\sigma^{\pm} = -i\frac{N_h ec}{H}q^2\,\Phi_{\pm}(q), \qquad (6.63)$$

where

$$\Phi_{\pm}(q) = \frac{1}{q^2}\sum_{s=-\infty}^{\infty}\frac{A_s^{\pm}}{4s\pm 1 - i\gamma}\mathcal{F}\left(\frac{q}{4s\pm 1 - i\gamma}\right), \qquad (6.64)$$

$s = 0, \pm 1, \pm 2, \dots$, and the parameter of nonlocality $q$ is the ratio of $u_{min}$ (the displacement of the holes at the minimum of $|\partial S/\partial k_z|$ during the cyclotron period) to the wavelength of the electromagnetic wave:

$$q = \frac{u_{min}}{\lambda}, \quad \text{where} \quad u_{min} = \frac{hcy_m}{|e|aH}. \qquad (6.65)$$

Note that $y_m = 8.1$ is a parameter of the model. The imaginary parts of $\Phi_-$ and $\Phi_+$ are identical, while the real parts differ only in sign, so it is sufficient to discuss only one of these functions.

In the collisionless limit the real part of $\Phi_+$, which we denote by $\text{Re}(\Phi_+^0)$, has two series of anomalies (infinities, in fact) at odd integer values of $q = m$ and at $q = M/\eta$ ($m, M$ are odd integers and $\eta = 1.1$ is the ratio of the maximum to the minimum magnitude of $|\partial S/\partial k_z|$). These anomalies are due to multiple DSCRs of the two groups of effective charge carriers; the odd-integer harmonics are the result of the particular symmetry of the cyclotron orbit.

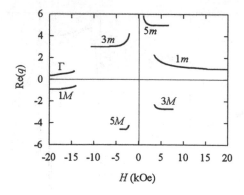

FIGURE 6.50. Dependence of $Re(q)$ on magnetic field calculated for $f = 200\,kHz$. Curve $\Gamma$ corresponds to a helicon. Curves $1m$, $3m$, and $5m$ correspond to multiple dopplerons for a minimum in $\partial S/\partial p_z$. Curves $1M$, $3M$, and $5M$ correspond to multiple dopplerons for a maximum in $\partial S/\partial p_z$. The dispersion curves shown for positive $H$ are associated with $(-)$ polarization, while the curves shown for negative $H$ are associated with $(+)$ polarization for $H > 0$. (After Fig. 3 in [120].)

The dispersion equation (4.99) may be written in terms of $\Phi_\pm$ as

$$\frac{\omega_0}{\omega} = \Phi_\pm(q), \tag{6.66}$$

where

$$\omega_0 = \frac{a^2 e H^3}{2\pi^2 h^2 N_h c y_m^2}. \tag{6.67}$$

Equation (6.66) has roots that correspond to waves propagating in both the positive and negative directions of the $z$ axis. Those traveling in the positive direction arise from roots that yield a positive group velocity ($\partial \omega/\partial k_z > 0$). Solutions relating to dopplerons caused by a maximum in $|\partial S/\partial k_z|$) have negative $q$ [or $(k_z)_D$] and, therefore, phase velocity, whereas those caused by a minimum, like a helicon, have a phase velocity of the same sign (or, keeping in mind that they are vectors) the same direction as the group velocity. Moreover, roots obtained for positive $Re(\Phi_0^+) > Im(\Phi_0^+)$ correspond to weakly damped waves of $(+)$ polarization, while those obtained for positive $Re(\Phi_0^-) > Im(\Phi_0^-)$ correspond to weakly damped waves of $(-)$ polarization for $H > 0$.

It can be seen that dopplerons of the same type but relating to different numbers of multiple resonances alternate their polarizations: $(+)$ for $(-)$ and vice versa. Solutions of equation (6.66) in the vicinity of resonance were obtained in the linear approximation of small parameters. They are given in Fig. 6.50.

Multiple DPRs have been investigated [37, 198] in indium specimens with $RRR \simeq 80\,000$ in the frequency range of 19.6 to 100\,MHz at $T = 4.2\,K$ and

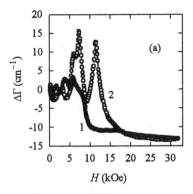 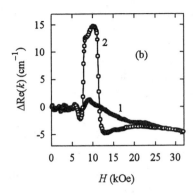

FIGURE 6.51. Magnetic field dependences of (a) absorption and (b) real components of the wave number for circularly polarized elastic-like modes in indium with [001] $\parallel$ **H** $\parallel$ **k**. Curve 1 is for $(-)$ polarization, Curve 2 for $(+)$. $f = 75\,\text{MHz}$, $T = 4.2\,\text{K}$, $\Delta\Gamma^{\pm} = \Gamma^{\pm}(H) - \Gamma^{\pm}(0)$, $\Delta k^{\pm} = k_e^{\pm}(H) - k_0 \sim \Delta s^{\pm}$. (After Figs. 3.10 and 3.11 in [198]. )

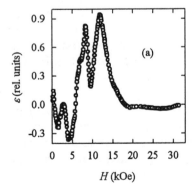 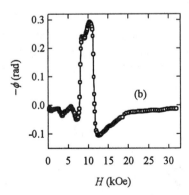

FIGURE 6.52. Magnetic field dependences of (a) ellipticity and (b) rotation of the polarization of ultrasound in indium. $l = 1.5\,\text{mm}$, the remaining parameters the same as in Fig. 6.51. (After Figs. 3.8 and 3.9 in [198].)

1.75 K. Figures 6.51(a) and 6.52(a) reveal two resonances that dominate over others. According to the positions of the absorption peaks they are observed at $H = 7.8\,\text{kOe}$ and $H = 11.3\,\text{kOe}$.

These peaks shift along $H$ when the frequency increases, but their shifts are nonlinear and different. The first one (which is at a lower field) exhibits a shift that is slower than linear, while the second is faster than linear. Such features should be typical for resonances due to the interaction of ultrasound with dopplerons of different types.

Recall that these interactions take place in the vicinity of the crossing of the dispersion curves for pure eigen modes: ultrasonic, electromagnetic, spin, etc. Therefore, in discussing DPR and the propagation of ultrasound along the positive direction of the $z$ axis, one should consider positive $(k_z)_e$ as well as $(k_z)_D$. In this case the usual dopplerons (caused by a maximum of $|\partial S/\partial p_z|$) have a negative group velocity. However, as was mentioned above, there exists another type of doppleron (caused by a minimum of $|\partial S/\partial p_z|$) that has phase and group velocities in the same direction.

Since the interaction of the waves results in coupled modes that reflect the properties of the initial waves, one should expect different manifestations of DPR caused by these two different types of dopplerons. Interaction with an ordinary doppleron (with strong coupling) forms a frequency gap for a given $H$ or magnetic field gap for a given $\omega$ (see Fig. 4.9): The dispersion equation has no solution in some interval of $\omega$ or $H$. Another type of doppleron having phase and group velocities of the same sign causes a gap in the dispersion curve similar to that caused by a helicon: At any $H$ and $\omega$ there are solutions relating to eigen modes (coupled in the vicinity of the resonance, and pure far from it).

When there is weak coupling, the elastic-like mode has an inverse $N$-shaped anomaly in the magnetic field dependence of the wave number (see Fig. 4.5). The dispersion curve of the elastic-like mode under DPR caused by a doppleron with identically directed phase and group velocities has a similar anomaly. On the other hand, a DPR caused by a doppleron with oppositely directed $\mathbf{v}_E$ and $\mathbf{g}_E$ should have an $N$-shaped anomaly in $k_z(H)$.

These two circumstances [nonlinear but different displacement of the resonant field versus frequency and existence of two different shapes of the resonant curves $\text{Re}[k_e(H)]$] make it possible to interpret the observed principal DPRs in indium as ones caused by dopplerons of different types and to explain the increase in $\Delta\text{Re}[k^+(H)]$ [Fig. 6.51(b)] and $\phi(H)$ [Fig. 6.52(b)] as a partial overlapping of two resonances.

Unfortunately, it was impossible to identify which electrons or holes are responsible for them since the dependences of the resonant fields on frequency were far from being close to their asymptotes and increasing frequency led to an absorption of ultrasound too high to conduct the measurements.

# 6.4   Helicon–Phonon Resonance

As noted above, HPR was discovered in electromagnetic experiments by Grimes and Buchsbaum [27], while Blaney [128] was the first to use the ultrasonic technique to investigate this phenomenon.

In certain metals (namely, those for which the plane $p_H = $ const produces a simply connected cross section of the FS, such as the noble metals) the only absorption mechanism for ultrasonic and helicon waves at fields above the Kjeldaas edge is that related to collisions. At liquid helium temperatures in pure metals the relaxation time can be large enough to satisfy the strong coupling condition, which in this case is given by equation (4.90).

This circumstance leads to a principal difference between the experimental observation of DPR and HPR. The first was studied exclusively under conditions of weak coupling between the elastic and electromagnetic subsystems. In this case two circularly polarized modes [one $(+)$ and one $(-)$] were generated by a piezoelectric transducer. As for HPR, it was observed under strong coupling and the signal registered in ultrasonic experiments was formed by the contributions of three circularly polarized modes: two coupled waves of one polarization and one pure elastic wave of the other circular polarization.

In turn, these differences in the signal structure resulted in dramatic differences in the magnetic field dependence of the ellipticity and the rotation of the polarization. Under weak coupling, the behavior of the ellipticity is similar (to some extent) to that of the absorption of the resonant elastic-like mode, and the rotation of the polarization is similar to a plot of the phase velocity [see equations (4.102) and (4.103)]. On the other hand, for strong coupling [this time equation (4.109) should be used] $\varepsilon$ and $\phi$ are determined by the interference between the coupled modes of resonant polarization and this interference provides minima and maxima in the signal amplitude and its phase shift for the resonant circular component, even in a nondissipative medium.

Both the ellipticity and the rotation of the polarization for strong coupling can be determined by the dispersion of the modes, but not by the absorption, since it takes place with weak coupling. Certainly, in reality one always observes a loss of wave energy that manifests itself by smoothing the anomalies in $\varepsilon(H)$ and $\phi(H)$. Although reducing the relaxation time (which can be done by warming the sample) can settle the regime of weak coupling, all of the experiments on HPR were performed in the strong coupling regime.

Thus, a review of the papers on experimental investigations of HPR in fact updates the consideration of MPP caused by a resonant interaction between the electromagnetic and elastic subsystems of a metal with a quite new situation which has never been observed in experiments on DPR.

Actually, HPR has been studied in two metals by means of the ultrasonic technique: potassium and indium. In potassium [128] the resonance was

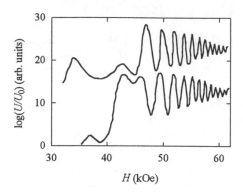

FIGURE 6.53. The variation with longitudinal magnetic field of the signal (on a logarithmic scale) detected by an AC-cut quartz transducer for 205.5 MHz waves propagating through a [100]-oriented potassium specimen. The two curves correspond to opposite directions of wave propagation with respect to the field. $T = 4.2$ K. (After Fig. 2 in [128].)

found for ultrasonic wave propagation along **H** and two crystallographic directions, [100] and [110]. Since polarization phenomena exist only along high-symmetry (not less than three-fold) axes, our particular interest is in the geometry involving [100]. The transmitting and receiving transducers were attached to the specimens with an angle of 30° between their polarization directions in order to register the effect of rotation of the polarization. The resonant field calculated for a frequency of 205.5 MHz in the free-electron approximation is expected to be about 70 kOe, which was beyond the limit of the magnet used (60 kOe). Records of the signal amplitude are shown in Fig. 6.53.

The observed oscillations were explained by the author as due to the rotation of the polarization of the linearly polarized sound waves. It was proposed that the linearly polarized wave was resolved into two circularly polarized modes of opposite polarization; one suffered no coupling with the helicon, while the other did interact with it and was changed in wavenumber. Introducing $\Delta k$ as the difference between the wavenumbers for the two opposite polarizations and $L$ as the length of the specimen, it was stated that a complete oscillation in the received signal occurred when $\Delta k L$ changed by $2\pi$. The author proposed that the finite height of the oscillations was due to the much stronger attenuation of the resonant mode because of its mixture with the helicon. A slight displacement of one curve with respect to the other was explained as due to the angle of 30° between the polarizations of the transducers: In one case, the initial rotation of the polarization was toward the polarization direction of the receiver, while for propagation in the opposite direction, the initial rotation was away from the receiver's polarization.

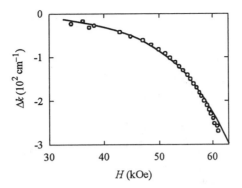

FIGURE 6.54. The variation with magnetic field of the experimental (open circles) and theoretical (solid line) values for the difference $\Delta k$ between the wavenumbers of the $(-)$ and $(+)$ polarized ultrasonic waves propagating along the [100] axis in potassium. (After Fig. 3 in [128].)

Figure 6.54 gives the theoretical curve for $\Delta k$ as a function of field with the appropriate parameters inserted. It was computed by solving equation (4.87) and, obviously, taking into account the elastic-like mode in the range below the resonant field. The points near the line are derived from the experiment. The zero of $\Delta k$ from the experimental results was only defined to within an additive factor $2\pi n/L$, where $n$ is an integer chosen to give the best fit to the theoretical curve.

The discrepancy between the experimental data and the theoretical curve was attributed to the need to take into account more accurately the finite value of the electron free path, but this statement as well as the use of only one mode of resonant polarization in interpreting the experimental data can be criticized.

First of all, with strong coupling both the phase shift as well as the amplitude of the signal associated with resonant circular polarization are the result of the interference of two coupled modes. One should use an approach such as that given in Sec. 4.5.2. Second, HPR was observed at a rather high magnetic field where the quasi-static elastic moduli can exhibit a field dependence that needs to be accounted for, even for wavenumbers of the mode which doesn't interact with the helicon. Thus the conditions for maxima and minima of the interference curve are much more complicated in reality than what was used in obtaining the data given in Fig. 6.54.

However, the good agreement between the experimental data and the calculated curve shows that the second resonant mode had a rather small amplitude, at least in the low-field region, and that the magnetic field dependence of the quasi-static elastic moduli contributes identically to all of the normal modes.

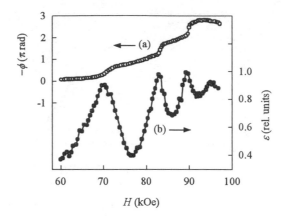

FIGURE 6.55. Magnetic field dependences of (a) the rotation of the polarization and (b) the ellipticity of ultrasound in a single crystal of indium with $\mathbf{k} \parallel \mathbf{H} \parallel [001]$, $f = 125\,\mathrm{MHz}$, $T = 2\,\mathrm{K}$, and $l = 2\,\mathrm{mm}$. (After Fig. 2 in [69].)

Another investigation of HPR that should be discussed here was performed in indium [69] where polarization phenomena were properly accounted for by using the technique presented in Chapter 2 [namely, equations (2.7), (2.21), and (2.22)].

Since indium has a tetragonal crystal structure, the Faraday effect can be observed only when the wave propagates along the [001] axis. The specimen had RRR $\cong 8 \times 10^4$ and a dimension along the wave path of 2 mm. The measurements were carried out in magnetic fields up to 97 kOe. The results of the experiments are presented in Fig. 6.55.

The resonance field calculated for a frequency of 125 MHz using equation (4.91) should be 91.8 kOe. Therefore, it was possible not only to investigate the low-field range but reach the resonance itself, contrary to what was done in [128] where the resonance field was beyond the limit of the magnet. All of the anomalies in the curves for $\varepsilon(H)$ and $\phi(H)$ in Fig. 6.55 shifted toward higher fields when the frequency was slightly reduced. This confirms their origin as HPR rather than DSCR or DPR. So, as was found for potassium [128], HPR in indium covers a rather broad range of magnetic fields. However, the curves obtained had such unusual and asymmetric shapes that they could not be analyzed in a general way.

To understand what can be interpreted in the framework of the existing theory, Gudkov and Zhevstovskikh [199] performed calculations of $\varepsilon(H)$ and $\phi(H)$ for helicon dispersion in the local regime using the parameters of indium. The circular components of the elastic displacements were written in the form of (4.107) and (4.108):

$$u^+(z) = u_0 \left( a_1 e^{-ik_1^+ z} + a_2 e^{-ik_2^+ z} \right),$$

$$u^-(z) = u_0 a_1 e^{-ik_s^+ z}, \tag{6.68}$$

where $a_1$ and $a_2$ describe the excitation efficiency of the coupled modes; their sum must satisfy

$$a_1 + a_2 = 1. \tag{6.69}$$

This is due to the assumption that a piezoelectric transducer generates linearly polarized vibrations. The ratio of the amplitudes of the resonant modes [(+) for indium] was found by using equation (4.110), while that for the transmission coefficients $d_{uu}^1$ and $d_{uu}^2$ was taken from [54]:

$$\frac{a_1^+}{a_2^+} = -\frac{k_1^+ \left( k_s^{+2} - k_2^{+2} \right)}{k_2^+ \left( k_s^{+2} - k_1^{+2} \right)} \equiv d, \tag{6.70}$$

where $k_s^{+2}$ is given by equation (4.78) and $k_{1,2}^{+2}$ by equation (4.89). It follows from these equations that

$$a_1^+ = \frac{1}{1 + d^{-1}} \quad \text{and} \quad a_2^+ = \frac{1}{1 + d}. \tag{6.71}$$

In order to derive the most important features of the resonance, it is useful to consider the region of magnetic field near $H_r$ [where it can be assumed that $a_1^+ = a_2^+ = 0.5$ and $\text{Im}(k_{1,2}^+) = 0$]. For this case

$$\varepsilon = \frac{\left| \cos \left( \frac{k_1^+ - k_2^+}{2} z \right) \right| - 1}{\left| \cos \left( \frac{k_1^+ - k_2^+}{2} z \right) \right| + 1}. \tag{6.72}$$

The extrema of this function are determined by the derivative

$$\frac{\partial \varepsilon}{\partial H} = \frac{\partial F}{\partial y} \frac{\partial y}{\partial H}, \quad \text{where} \quad y \equiv \frac{k_1^+ - k_2^+}{2} z, \tag{6.73}$$

so that

$$\frac{\partial F}{\partial y} = -\frac{2 \sin y}{(1 + \cos y)^2}, \quad (\cos y > 0),$$

$$\frac{\partial F}{\partial y} = -\frac{2 \sin y}{(1 - \cos y)^2}, \quad (\cos y < 0). \tag{6.74}$$

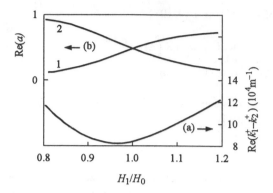

FIGURE 6.56. Magnetic field dependence of (a) $\mathrm{Re}(k_1^+ - k_2^+)$ and (b) $\mathrm{Re}(a_1^+)$ (Curve 1) and $\mathrm{Re}(a_2^+)$ (Curve 2) calculated in the local regime using the properties of indium and $\tau = 10^{-8}\,\mathrm{s}$, $f = 125\,\mathrm{MHz}$. (After Fig. 4 in [199].)

Points where $\partial F/\partial y = 0$ and has discontinuities are related to the minima and maxima of $|\varepsilon|$ with magnitudes equal to 0 and 1. The zero of $\partial y/\partial H$ gives an extremum at $H = H_1$ that corresponds to a minimum of the difference $k_1^+ - k_2^+$ and, as can be seen in Fig. 6.56, this field is less than $H_r$. The difference between $H_1$ and $H_r$ seems to be the result of the nonlinear dispersion curves (or at least one of them) corresponding to those calculated when the interaction between the subsystems of a metal was neglected; in our case the helicon has nonlinear dispersion.

In order to understand what type of extremum will be observed at $H_1$, the ellipticity of ultrasound was calculated for a metallic specimen with the parameters of indium. Figure 6.57 shows that $|\varepsilon|$ can display any type of extremum, either a maximum or a minimum, depending upon the distance covered by the ultrasound in the specimen.

Using the magnetic field dependence of the amplitudes of the resonant modes [as given for $a_1(H)$ and $a_2(H)$ in equation (6.71)] yielded the plot shown in Fig. 6.58(a). It can be seen that taking into account the differences in the amplitudes of the waves generated by the transducer leads to reduction of the maximum values of $|\varepsilon|$, while the minima are unchanged and equal to zero. The first is because of the fact that the components $a_1 e^{-ik_1^+}$ and $a_2 e^{-ik_2^+}$ do not cancel completely, even for fields corresponding to $(k_1^+ - k_2^+)z = \pi(2n+1)$; the second is due to the vanishing of $u^+(z)$ at $(k_1^+ - k_2^+)z = 2\pi n$, where $n = 0, \pm 1, \pm 2, \ldots$.

The difference in the amplitudes of the modes of resonant polarization results in oscillations of $\phi(H)$. Such oscillations calculated with the use of equations (2.7), (4.78), (4.89), and (6.68)–(6.71) are given in Fig. 6.58(b). For convenience the principal values of $\phi(H)$ are given $(-\pi/2 < \phi \le \pi/2)$.

An analysis of the influence of losses on $\varepsilon(H)$ shows that it depends on the subsystem in which the energy losses dominate, the elastic or the electro-

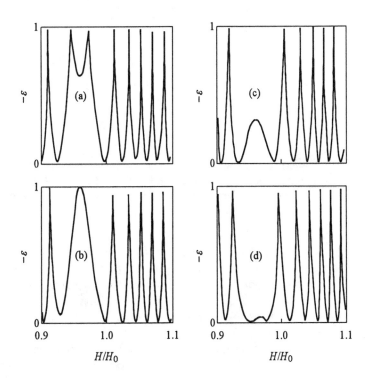

FIGURE 6.57. Magnetic field dependence of the ellipticity of ultrasound calculated for $a_1 = a_2$ and $\alpha^{\pm} = 0$. (a) $l = 2.020\,\mathrm{mm}$, (b) $l = 2.026\,\mathrm{mm}$, (c) $l = 2.040\,\mathrm{mm}$, and (d) $l = 2.060\,\mathrm{mm}$. The other parameters are the same as in Fig. 6.56. (After Fig. 3 in [199].)

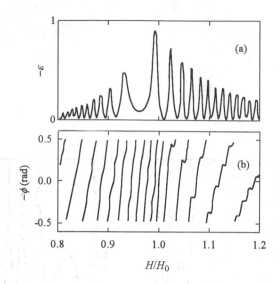

FIGURE 6.58. Magnetic field dependence of (a) the ellipticity and (b) the rotation of ultrasound, calculated for $a_1$ and $a_2$ as given by (6.71). Here $l = 2\,\mathrm{mm}$, $\alpha^{\pm} = 0$, and the other parameters are the same as in Fig. 6.56. (After Figs. 5 and 6 in [199].)

magnetic one. Figures 6.59 and 6.60 show the ellipticity and the imaginary components (that determine the absorption) of the wave numbers relating to the resonant polarization modes. The curves in (a) correspond to approximately equal losses in both systems in the vicinity of the resonance, in (b) the dominant losses are in the electromagnetic subsystem, and in (c) they are in the elastic one. In carrying out the calculations, the absorption in the elastic subsystem was considered to be independent of $H$ and was described by the real part of $\alpha^{\pm}$ that corresponds to collisionless deformation absorption.

Taking losses into account, compared with taking $a_1(H)$ and $a_2(H)$ into account, results primarily in changing the values of $\varepsilon(H)$ at its minima. This is more apparent in Curves (b) and (c) of Fig. 6.59, where the latter even alternates in sign. At the same time, in all of the curves there is a large oscillatory contribution for fields above $H_r$ owing to the increase of the coupling parameter at high fields.

In summary, it should be noted that investigations of HPR in indium have revealed some specific features and showed new opportunities for the investigation of the electronic properties of a metal by measuring the polarization of ultrasound. Specifically:

1. Oscillatory dependences of the ellipticity and rotation of the polarization of ultrasound have been observed in experiments and in model calculations for a wide range of magnetic fields. The wide range over which HPR

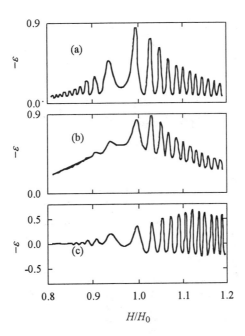

FIGURE 6.59. Magnetic field dependence of the ellipticity of ultrasound calculated for a metal with the properties of indium and $l = 2\,\text{mm}$, $f = 125\,\text{MHz}$. (a) $\tau = 10^{-9}\,\text{s}$, $\alpha^{\pm} = 10^{-3}\,\text{N}\,\text{s}/\text{m}^2$, (b) $\tau = 10^{-10}\,\text{s}$, $\alpha^{\pm} = 0.03\,\text{N}\,\text{s}/\text{m}^2$, and (c) $\tau = 10^{-10}\,\text{s}$, $\alpha^{\pm} = 0.055\,\text{N}\,\text{s}/\text{m}^2$. (After Fig. 7 in [199].)

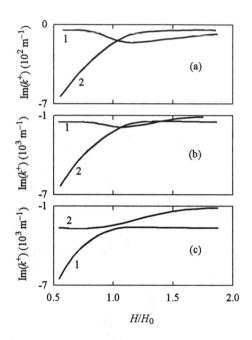

FIGURE 6.60. Magnetic field dependence of the imaginary components of the wavenumbers for the resonant polarization modes. Curves labeled 1 correspond to $k_1^+$, those labeled 2 to $k_2^+$. The other parameters are the same as in Fig. 6.59. (After Fig. 8 in [199].)

is manifested is one of the most important features of the resonance that is observed when there is strong coupling between the elastic and, in this case, electromagnetic subsystems. This fact was noticed also in the very first investigation by Blaney [128]. This range of resonance anomalies in $\varepsilon$ and $\phi$ is due to the absence of strong $H$ dependences of the amplitudes of the resonant coupled waves and relatively small variations of the dispersion curves in the vicinity of $H_r$.

2. The frequency of the $\varepsilon$ and $\phi$ oscillations measured as functions of $H$ is approximately 1.5 times larger than in the theoretical curves. The difference could be due to the use of the local approximation for the conductivity that was used in the calculations but, in reality, nonlocal effects (DSCR) can influence $\sigma^{\pm}(H)$ and, therefore, the dispersion of the coupled modes, particularly for $H < H_r$.

3. The amplitude of the $\varepsilon$ oscillations is reduced for an increase in the magnetic field in the experiments, while the calculated curves manifest the opposite variation. This proves that the coupling parameter for $H < H_r$ exceeds that for $H > H_r$. This fact also can be explained by the influence of DSCR on the coupling parameter, which has resonance anomalies near the Kjeldaas edge. In its most general form the parameter should be written as the right-hand side of equation (4.77).

4. Model calculations have shown that an alternating-sign behavior of $\varepsilon(H)$ can be obtained only if the energy losses in the elastic subsystem dominate those in the electromagnetic subsystem. Experimental data show that the ellipticity has a constant-sign behavior with respect to $H$. However, it should be pointed out that in indium the energy losses in the electromagnetic subsystem are dominant or at least of the same order as in the elastic subsystem.

## 6.5   Giant Geometric Oscillations

Another possible manifestation of magnetoacoustic geometric oscillations has the form of anharmonic oscillations. In contrast with the situation discussed in the previous section, these oscillations occur when the electromagnetic term dominates in equation (6.9) for the wave vector of the elastic-like mode.

One possible reason for the appearance of anharmonic oscillations is the fact that the electroacoustic coefficient $\beta^{\pm}$ is squared in this expression. Since the coefficient itself oscillates harmonically as a function of inverse magnetic field, the absorption oscillations caused by the electromagnetic mechanism should contain a doubled-frequency component. Their combination with oscillations originating from the deformation term leads to complicated variations of $\Gamma^{\pm}(H)$ and $s^{\pm}(H)$ and, therefore, of $\varepsilon(H)$ and $\phi(H)$.

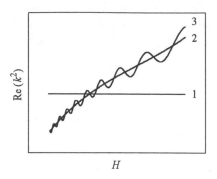

FIGURE 6.61. The magnetic field dependence of the squared wavenumbers for (1) an ultrasonic mode, (2) a doppleron in the absence of geometric oscillations, and (3) a doppleron along with geometric oscillations. (After Fig. 2 in [200].)

A second possible reason is an interaction between the ultrasonic wave and dopplerons. Recall that the dispersion of dopplerons is determined by the circular components of the nonlocal conductivity, $\sigma^{\pm} = \sigma_{11} \pm i\sigma_{21}$, which is a function of the wave vector and the magnetic field,

$$i\frac{(ck_D^{\pm})^2}{4\pi} = \omega\sigma^{\pm}, \qquad (6.75)$$

where $k_D^{\pm}$ is the wavenumber of the doppleron. When $|\sigma_{21}| \gg |\sigma_{11}|$, equation (6.75) has a solution corresponding to a weakly damped mode. It can exist for only one circular polarization at a given magnetic field.

Upon replacing $\omega\sigma^{\pm}$ by $i(ck_D^{\pm})^2/4\pi$, one sees that the denominator of equation (6.9) takes into account only the interaction between the acoustic $(k_0)$ and the electromagnetic $(k_D^{\pm})$ waves. Actually, this is the most significant interaction. It has a multi-resonance character with the resonances determined by the intersections of the dispersion curves for the ultrasonic wave and the doppleron. Figure 6.61 shows a schematic representation of this condition for constant frequency.

If geometric oscillations are absent, DPR manifests itself in an ultrasonic experiment as an absorption peak accompanied by variations of the phase velocity of the circularly polarized wave. We have discussed this case in Sec. 6.3.

In discussing the possible influence of DPR on the geometric oscillations of ultrasonic absorption and phase velocity, we must note the following. When the oscillations of $k^{\pm}$ are mainly due to the oscillating component of $\beta^{\pm}$, DPR should increase the amplitude of the oscillations. Of course, this is true if their period is smaller than the linewidth of the resonance with respect to $H$. This case is similar to the amplification of quantum oscillations caused by the interaction of ultrasound with a helical wave that is discussed in Sec. 6.6.

However, a more complicated variation of the shape of the first kind of oscillation may occur if the Hall component of the conductivity undergoes significant geometric oscillations. In this case the wave vector of the doppleron also has an oscillatory part, so the system periodically approaches and departs from the resonance with monotonic variation of $H$. As a result, the interaction of ultrasound with electromagnetic waves not only can amplify geometric oscillations, raising the $\Gamma^{\pm}$ amplitudes up to a level comparable with the total electronic absorption, but it also may change their shape significantly, such as causing the appearance of a minimum at the position of a former maximum or even new intersections of the dispersion curves that become in fact new resonances. Taking these circumstances into account, Gudkov and Tkach [172] suggested naming them *giant geometric oscillations* (GGO), in analogy with giant quantum oscillations [147]. Since DPR is observable for one of the circular polarizations, the characteristic feature of giant geometric oscillations of ultrasonic absorption is the considerable discrepancy of their amplitude from that for the alternative polarization.

Giant geometric oscillations of ultrasonic absorption and ellipticity were found in indium by Gudkov and Tkach [200–202]. The measurements were performed in ultrapure $10 \times 10 \times 2 \, \mathrm{mm}^3$ single-crystal specimens which had $\omega\tau \approx 1$ at a frequency of 63 MHz and $T = 1.28$ K. The propagation direction for the circularly polarized ultrasonic waves coincided with the tetragonal crystallographic axis to within an accuracy of $4°$.

For magnetic fields less than 500 Oe, relatively small oscillations of absorption and phase velocity were observed for both of the circular polarizations with approximately equal amplitudes. In this region oscillatory components of the electroacoustic coefficients are small and oscillations of $\Gamma^{\pm}$, $s^{\pm}$, $\varepsilon$, and $\phi$ are similar to the harmonic ones that were discussed in Sec. 6.2.

For larger values of $H$, significant discrepancies between the absorption curves for the different polarizations begin to appear. Appreciable anharmonic distortions emerge and the amplitude for the $(-)$ polarization exceeds that for the $(+)$. The behavior of $\Gamma^{\pm}(H)$, measured relative to the absorption level in the superconducting state with $H = 0$, is shown in Fig. 6.62. The differences between the $\Gamma^+$ and $\Gamma^-$ oscillations give rise to oscillations of the ellipticity as shown in Fig. 6.63.

The gross behavior of $\Gamma^{\pm}(H)$ in indium conforms quite well to the characteristic traits of the anharmonic geometric oscillations discussed above, and to the giant ones in particular. At the same time, a detailed interpretation of the experimental data is significantly impeded by the superposition of several different magnetoacoustic effects, including geometric oscillations, DSCR, and multiple DPRs, caused by different groups of charge carriers [37].

Let us try to understand qualitatively the main factors leading to the observed shapes of the oscillations. One circumstance that facilitates the

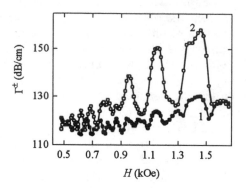

FIGURE 6.62. Absorption of circularly polarized ultrasonic waves in ultrapure indium for [001] ∥ **H** ∥ **k**: (1) (+) polarization, (2) (−) polarization. $f = 63.2$ MHz, $T = 1.28$ K. (After Fig. 1 in [200].)

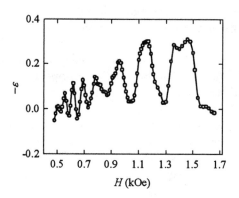

FIGURE 6.63. Ellipticity of ultrasound in an ultrapure indium crystal for $l = 2$ mm. The remaining parameters are the same as in Fig. 6.62. (After Fig. 2 in [201, 202].)

interpretation is the following: The effects due to DSCR and DPR should not depend strongly on temperature in the interval $1.2 < T < 4.2\,\mathrm{K}$ for indium crystals of such perfection. The reason is that the terms of the electroacoustic coefficients [with $m \neq 0$ in equation (6.12)] describing the above-mentioned effects are determined by the values of $k_0 l$ and $\Omega \tau$, which significantly exceed unity in our experiment (here $l$ is the electron free path). As a consequence, the $H$-dependences of these terms are close to their limiting values.

On the other hand, geometric oscillations of an appreciable amplitude appear only when $\omega \tau$ becomes comparable with unity [see equation (6.23) and the associated comments], i.e., for temperatures near $1\,\mathrm{K}$ in the specimens investigated. It may be concluded from an analysis of the curves for $\Gamma^{\pm}(H)$ in [37] obtained at $T = 4.2\,\mathrm{K}$ that, in the field region $0.9 < H < 1.6\,\mathrm{kOe}$, two overlapping peaks can be observed for $(-)$ polarization that were due to multiple DPRs. Only in this field region is an increase in the amplitude of the geometric oscillations recorded for $T = 1.28\,\mathrm{K}$, predominantly for the resonance polarization (see Fig. 6.62). The increase cannot be explained by the linear magnetic field dependence of the amplitude of the harmonic geometric oscillations: when the magnetic field was doubled, the oscillation amplitude for the $(-)$ polarization grew more than three times. At the same time, the amplitude variation for the other polarization can be described with the help of a nearly linear dependence. In addition, for both polarizations doubled-frequency components appeared, which should be caused by the $\beta^{\pm}$ contribution.

Thus, in the vicinity of DPR, harmonic oscillations of the electroacoustic coefficients can result in anharmonic giant oscillations of absorption and ellipticity of ultrasound. As for oscillations of phase velocity and rotation of polarization, it should be mention that it is probably possible to observe anharmonic oscillations of these properties but, since the electronic contribution to the phase velocity is small, these geometric oscillations will never be "giant," at least as far as the phase velocity is concerned.

# 6.6  Quantum Oscillations

Everything that has been said up to this point about nonmagnetic metals refers to the quasi-classical approach to the problem. However, at low temperatures and high magnetic fields a new group of effects appear, which are united by the common term *quantum oscillatory phenomena*. They are caused by the quantization of the energies of conduction electrons in a sufficiently strong magnetic field (Landau quantization [203]) or, more precisely, of the energies of their motion in the plane perpendicular to **H**.

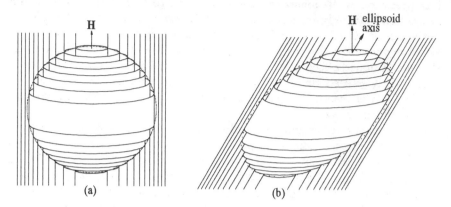

FIGURE 6.64. Landau tubes for (a) a spherical and (b) an elliptical Fermi surface. (After Fig. 2.1 in [204].)

Thus, $\mathcal{E}_0$ in equation (4.64) should be written as

$$\mathcal{E}_0 = \left(n + \frac{1}{2}\right)\hbar\Omega + \frac{p_H^2}{2m}, \tag{6.76}$$

where $n$ is the integer that denotes the number of the energy sub-band, or the Landau level.

The geometric objects representing the regions in $\mathbf{p}$-space where electrons can exist are known as Landau tubes (see Fig. 6.64). The condition $\mathcal{E}_0 \leq \mathcal{E}_F$ denotes the occupied and empty states on these tubes at $T = 0$. Scattering of the electrons by thermal phonons, impurities, and crystal defects can destroy the discrete structure of the energy spectrum. Thus the conditions which are necessary in order for quantum effects to appear are

$$\Omega\tau \gg 1, \quad \hbar\Omega > k_B T_D, \tag{6.77}$$

where $T_D$ is the Dingle temperature (the parameter which accounts for the influence of both finite temperature and finite relaxation time).

Most of the properties of a metal are determined by the electrons on the FS. As the magnetic field increases the dimensions (in the plane perpendicular to $\mathbf{H}$) of the Landau tubes grow and, when a tube leaves the FS and therefore is no longer occupied by electrons, there should be an anomaly in the physical properties of the metal. These anomalies repeat periodically when plotted against inverse magnetic field and are called quantum oscillations. They were first observed in electrical conductivity (the Shubnikov–de Haas effect [205]) but very soon also in magnetic susceptibility (the de Haas–van Alphen effect [206]), and later in many other parameters determined directly or indirectly by the distribution function (4.57) for the conduction electrons. Quantum oscillations (QO) became a useful instrument for investigating the FS of metals since their period in inverse magnetic field $\Delta(1/H)$ is directly related tp the extremal cross-sectional area $S_{ext}$ of the FS in the plane perpendicular to **H**:

$$\Delta\frac{1}{H} = \frac{2\pi|e|\hbar}{cS_{ext}}. \tag{6.78}$$

A detailed description of the most important oscillatory phenomena in metals related to QO can be found in the book by Shoenberg [204].

The propagation parameters of ultrasonic waves also exhibit QO. First discovered by Reneker [207, 208] as oscillations of absorption, they proved to be of two different types:

1. The oscillations of the quasi-static elastic moduli found in bismuth by Mavroides et al. [209] are the first type and are analogous to the oscillations of thermodynamic quantities. They manifest themselves as small (in comparison with the monotonic part) oscillations in phase velocity, usually of a quasi-harmonic form if plotted against $1/H$. This kind of oscillation has been reviewed by Tesrardi and Condon [210].

2. The oscillations of the second type are giant quantum oscillations (GQO) of ultrasonic absorption. They were predicted by Gurevich et al. [147] and observed by Koroljuk and Pruschak in zinc [148]. These oscillations are of the order of the entire electronic absorption or even greater, and have the form of well-resolved peaks that are also periodic in inverse magnetic field. Oscillations of collisional absorption are also possible but they are usually less important and difficult to observe, like Shubnikov–de Haas oscillations [205]. Shapira has described GQO in detail in a review [211].

QO of polarization parameters were observed by Gudkov and Vlasov in tungsten under DPR [66] and by Gudkov and Zhevstovskikh [212] in high magnetic fields (with respect to the Kjeldaas edge, i.e., for $H \gg H_K$). It should be noted that QO of ellipticity and rotation of the polarization of ultrasound cannot be introduced as oscillations of a new physical parameter since $\varepsilon$ and $\phi$ are functions of absorption or phase velocity of the eigen modes that have elastic displacement as one of their characteristics. In view of this, QO of the polarization parameters arise when the oscillatory components in absorption or phase velocity are different for waves of different circular polarizations. Obviously, such cases are very rare and each requires individual theoretical study.

In the first paper [66] the oscillations (see Fig. 6.65) were investigated at comparatively small magnetic fields so that the condition for resolving the Landau levels under DPR was just beginning to be fulfilled. Thus, the QO were very small and had a nearly harmonic form as can be seen in Figs. 6.66 and 6.67. The period of the oscillations was $(0.116 \pm 0.002) \times 10^{-6}\,\text{Oe}^{-1}$ and, in accordance with the data obtained by Lee et al. [213] they were initiated by the holes located on the $N$-ellipsoids (see Fig. 6.68). Moreover, taking into account the symmetry of the cyclotron orbits of the holes in

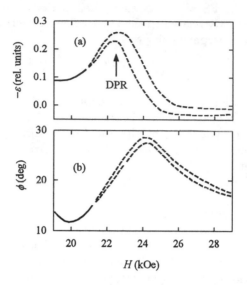

FIGURE 6.65. The magnetic field dependences of ellipticity and rotation of the polarization of ultrasound under DPR in tungsten with [001] ∥ **H** ∥ **k**, RRR $\geq 10^5$, $l = 4.57\,\text{mm}$, $f = 192\,\text{MHz}$, and $T = 1.8\,\text{K}$. Dashed lines indicate the envelope of quantum oscillations. (After Fig. 1 in [66].)

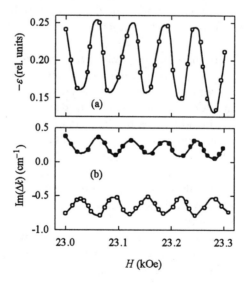

FIGURE 6.66. Expanded views showing quantum oscillations of (a) the ellipticity of ultrasound and (b) the imaginary components of the wave vectors of circularly polarized elastic-like modes [open circles, (+) polarization; filled circles, (−)]. The other parameters are as in Fig. 6.65. (After Fig. 2 in [66].)

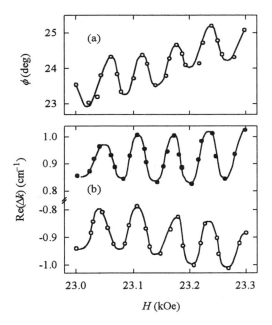

FIGURE 6.67. Expanded views showing quantum oscillations of (a) rotation of the polarization of ultrasound and (b) real components of the wave vectors of circularly polarized elastic-like modes [open circles, (+) polarization; filled circles, (−)]. The remaining parameters are as in Fig. 6.65. (After Fig. 3 in [66].)

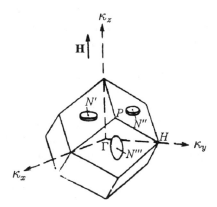

FIGURE 6.68. Positions of the hole $N$-ellipsoids with respect to the center of the tungsten Brillouin zone $\Gamma$. The heavy lines show the effective charge carriers which cause quantum oscillations in the ultrasonic parameters of tungsten for the field configuration shown. (After Fig. 3 in [214].)

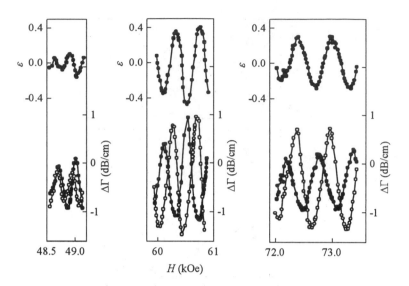

FIGURE 6.69. Absorption of circularly polarized elastic-like modes and ellipticity of ultrasound versus magnetic field in tungsten with $[001] \parallel \mathbf{H} \parallel \mathbf{k}$, $L = 4.57\,\mathrm{mm}$, $\mathrm{RRR} \geq 10^5$, $f = 196\,\mathrm{MHz}$, and $T = 1.8\,\mathrm{K}$. Open circles, $(+)$ polarization; filled circles, $(-)$. (After Fig. 1 in [214].)

the effective zone, it is possible to show that the $N$-ellipsoid terms in $\beta^{\pm}$ and $\sigma^{\pm}$ should be very small in the range of magnetic fields corresponding to DPR, whereas the terms in $\alpha^{\pm}$ differ from zero even for magnetic fields $H \gg H_K$ [181]. However, the deformation terms are identical for both of the polarizations in high fields and fail to explain the existence of QO in $\varepsilon$ and $\phi$.

In order to explain the existence of the oscillations the authors proposed that the effect can be described by solving equation (4.77) as done previously by an iterative method, but adding a second approximation: inserting $k^{\pm}$ instead of $k_0$ in the right side of equation (4.92). The most substantial result of this procedure is the appearance of an oscillatory imaginary component of $k^{\pm}$ (obtained at the first step of iteration) in the denominator of the field term. Thus, the deformation absorption related to the $N$-ellipsoid term in $\alpha^{\pm}$ gives rise to additional oscillatory differences in the dispersion of the circularly polarized elastic-like modes, resulting in the appearance of QO of ellipticity and rotation of the polarization, whereas the main effects in $\varepsilon(H)$ and $\phi(H)$ were due to the field term. This phenomenon reflects the well-known fact that any resonance anomaly depends on the energy dissipation in a system. However, detailed model calculations were not carried out to determine whether this interpretation was correct or not.

The observation of quantum oscillations of $\varepsilon$ (shown in Fig. 6.69) and $\phi$

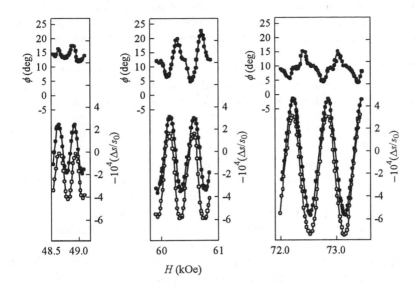

FIGURE 6.70. Phase velocity of circularly polarized elastic-like modes and rotation of the polarization of ultrasound versus the magnetic field in tungsten. The parameters are the same as in Fig. 6.69. (After Fig. 2 in [214].)

(Fig. 6.70) in high magnetic fields was rather unexpected and impossible to comprehend at first. The reason was that only the $\alpha^+ = \alpha^-$ components have a nonzero high-field limit, though the differences in phase velocities and absorption of the circularly polarized elastic-like modes should diminish far from DPR, resulting in vanishing ellipticity and rotation of the polarization, both the oscillatory and nonoscillatory parts.

An interpretation of the oscillations of the polarization parameters in high magnetic fields was presented by Bebenin et al. [215–217]. It involved the statements that the real and imaginary parts of the electroacoustic coefficients do not reach their high-field limits in the field range investigated, both types of quantum oscillations take part in forming the shape of the oscillations and, in addition, there is a resonant interaction with a helical electromagnetic wave in a compensated metal predicted by Kaner and Skobov [111] that amplifies the oscillations of phase velocity and absorption for one of the circular polarizations.

As has been mentioned already, QO of the period observed are due to the charge carriers in the central cross sections of the hole ellipsoids. The centers of the ellipsoids coincide with the centers of the rhombi ($N$-type points) that form the Brillouin zone. The axes are oriented along the $N\Gamma$, $NP$, and $NH$ directions. Figure 6.68 shows three ellipsoids; the nine others which have corresponding arrangements have been omitted to keep the figure simple. Ellipsoids with centers at the $N'$- and $N''$-type points (there

are eight such ellipsoids) contribute to the oscillations. The four other ellipsoids, with centers at the $N'''$-type points, which lie in the plane perpendicular to $H$ and passing through the center of the Brillouin zone $\Gamma$, do
not contribute to the oscillations.

The solution of the dispersion equation for the elastic-like modes can be
written in the form of the quasi-classical equation (4.92), even taking into
consideration quantum effects. The difference between the quasi-classical
and the quantum approaches lies only in the form of the electroacoustic
coefficients that are presented as a sum of quasi-classical ones and small
corrections that account for quantum effects.

The quasi-classical electroacoustic coefficients, given in their general form
by equations (4.74), (4.80), and (4.81), can be expanded in a series in
$\mathbf{k}_0 \cdot \bar{\mathbf{v}}_{max}/\Omega$ for $H \gg H_K$. Here the parameter of nonlocality $q$ is the
same as that initially introduced in equation (4.96), but now the effective
charge carriers become those having the greatest value of velocity averaged
over the cyclotron period, although $q = 1$ defines the Kjeldaas edge at a
given frequency. The quasi-classical coefficients $\alpha^{\pm}$ and $\beta^{\pm}$ in the high-field
approximation depend on the off-diagonal components of the deformation
potential averaged over the cyclotron orbit. Such an averaged component
relating to the $N'$- or $N''$-ellipsoid in the coordinate system corresponding
to axes of the principal ellipsoid is written as $\Lambda_e = (\Lambda_2 - \Lambda_3)/2$, where
$\Lambda_2$ and $\Lambda_3$ are the principal values of the deformation potential tensor. In
addition $\beta^{\pm}$ is proportional to an off-diagonal component of the momentum
flux tensor $\Pi_e = \mathcal{E}_F(m_3 - m_2)/(m_3 + m_2)$, where $\mathcal{E}_F$ is the modulus of the
Fermi energy measured with respect to the maximum in the dispersion law
at the $N$-point, and $m_i$ ($i = 1, 2, 3$) are the principal values of the effective
mass tensor in the same coordinate system.

As a result, the quantum corrections to the electroacoustic coefficients
can be written as

$$\alpha_q^{\pm} = \frac{\lambda_e}{i\omega}\left(K_{to} + i\frac{\pi s_0}{2v_F}K_{go}\right),$$

$$\beta_q^{\pm} = \frac{ick_0\lambda_e}{\omega H}\zeta\left(K_{to} + i\frac{\pi s_0}{2v_F}K_{go}\right), \qquad (6.79)$$

where $\lambda_e = 2g\Lambda_e^2$, $g = |m|m_{zz}v_F/\pi^2\hbar^3$ is the density of states at the
Fermi level on one ellipsoid, $v_F = \sqrt{\mathcal{E}_F/m_{zz}}$, $m = \sqrt{(m_1 m_2 m_3)/m_{zz}}$ is
the cyclotron mass, $m_{zz} = (m_2 + m_3)/2$ is the longitudinal effective mass,
and $\zeta = \Pi_e/\Lambda_e$. The functions $K$ and $K_g$ describe two types of quantum
oscillations, thermodynamic and giant, respectively:

$$K_{to} = \sqrt{\frac{\hbar\Omega_e}{2\mathcal{E}_F}}\sum_{n=1}^{\infty}\frac{(-1)^n n\Theta}{\sqrt{n}\sinh(n\Theta)}\cos\left(\pi n\frac{2\mathcal{E}_F}{\hbar\Omega_e} - \frac{\pi}{4}\right)\cos\left(\pi n\frac{\Omega_0}{\Omega_e}\right), \qquad (6.80)$$

$$K_{go} = 2 \sum_{n=1}^{\infty} \frac{(-1)^n n\Theta}{\sinh(n\Theta)} \cos\left(\pi n \frac{2\mathcal{E}_F}{\hbar\Omega_e}\right) \cos\left(\pi n \frac{\Omega_0}{\Omega_e}\right), \qquad (6.81)$$

where $\Theta = 2\pi^2 k_B T/\hbar\Omega_e$, $\Omega_e$ is the cyclotron frequency of the effective charge carriers (holes near the central cross-section of the ellipsoids in the $p_x - p_y$ plane), and $\hbar\Omega_0$ is the spin splitting of the Landau levels in a magnetic field. Equation (6.81) is valid for $k_0\ell \gg \sqrt{\mathcal{E}_F/k_B T}$. It is of great importance for our considerations that $K_{to}$ and $K_{go}$ have a phase shift of $\pi/4$.

The value of $\zeta$ is approximately $10^{-2}$, so if we restrict ourselves to the approximation that $k^{\pm}$ is linear in $\zeta$, we should ignore the quantum corrections to the conductivity $\sigma^{\pm}$. As for the quasi-classical expressions, they should be given as expansions in powers of $q$ (recall that the nonlocality parameter $q$ is proportional to $k/H$ and it is small for $H \gg H_K$),

$$\sigma_{cl}^{\pm} = i\frac{c^2 k^2}{4\pi}\left(\mp\frac{1}{\omega_s} - i\tau_s\right), \qquad (6.82)$$

$$\beta_{cl}^{\pm} = \pm\frac{ckH}{4\pi i\omega}\frac{\omega}{\omega_s}\Delta, \qquad (6.83)$$

where $\omega_s$ and $\tau_s$ are proportional to $H^3$ and $1/H^2 k^2$, respectively, with coefficients that depend on the shape of the FS, and the real dimensionless parameter $\Delta$ depends on the deformation potential tensor. Actually $\Delta$ is proportional to the $\Lambda_{13}$ component averaged over the FS taking into account all of the sheets, not just the ellipsoids. Expressions for the parameters $\omega_s$, $\tau_s$, and $\Delta$ have been derived in [216].

While discussing the dispersion equation (4.77) in Section 4.4, we mentioned that, when the coupling parameter is equal to zero, the equation transforms into two, which describe the dispersion of pure elastic and pure electromagnetic waves. The latter dispersion equation is

$$k^{\pm^2} - k^{\pm^2}_E = 0. \qquad (6.84)$$

Upon substituting equation (6.82) into (4.79), which defines $k_E^{\pm^2}$, and the result into equation (6.84), one can see that $\omega_s$ is the eigenfrequency of a helical wave. Kaner and Skobov [111] predicted such a wave in a compensated metal and, in fact, it relates to the long-wavelength part of the doppleron spectrum, but at that time DSCR-modes had been discovered only in noncompensated metals [107, 108].

Interactions with this wave made it possible to explain the fact that the observed magnitude of the quantum oscillations of $\varepsilon$ and $\phi$ are very small if one considers zero Hall conductivity as it should be in the high-field limit for a compensated metal. Note that this electromagnetic wave is related to a different branch of the spectrum than the $G$-doppleron discussed in

Sec. 6.3. In an ultrasonic experiment a resonant interaction of the elastic wave with a definite doppleron branch can take place only once (only one intersection of the two dispersion curves can exist at a given frequency) and the $G$-DPR was observed at lower fields near the Kjeldaas edge. Besides, this resonance was found for a different circular polarization than that of the $G$-doppleron.

Substitution of the quantum corrections and the electroacoustic coefficients in their quasi-classical form evaluated at $k = k_0$ into equation (4.92) gives for the oscillatory components of the wave vectors

$$\mathrm{Re}(\Delta k_{os}^{\pm}) = \frac{k_0}{2c_{1313}}\lambda_e\left[\left(1 + 2\zeta\Phi'_{\pm}\right)K_{to} - 2\zeta\frac{\pi s_0}{2v_F}\Phi''_{\pm}K_{go}\right], \qquad (6.85)$$

$$\mathrm{Im}(\Delta k_{os}^{\pm}) = \frac{k_0}{2c_{1313}}\lambda_e\left[2\zeta\Phi''_{\pm}K_{to} + \left(1 + 2\zeta\Phi'_{\pm}\right)\frac{\pi s_0}{2v_F}K_{go}\right], \qquad (6.86)$$

where $\Phi'_{\pm}$ and $\Phi''_{\pm}$ are the real and imaginary parts, respectively, of the function that accounts for the resonant interaction with the helical wave,

$$\Phi_{\pm} = \frac{1 \mp \omega\Delta/\omega_s}{1 \mp \omega/\omega_s - i\,\omega\tau_s}. \qquad (6.87)$$

Equations (6.85) and (6.86) show that both the real and the imaginary components of the wave numbers of the circularly polarized elastic-like modes contain two oscillatory functions, $K_{to}$ and $K_{go}$. These fuctions, according to (6.80) and (6.81), have different phase shifts. In (6.85) and (6.86) they are multiplied by coefficients that are different for each function and dependent on polarization. Therefore, in general one expects there to be differences in the amplitudes and phases of the oscillations of both absorption and phase velocity of the waves of different circular polarization that can result in oscillations of ellipticity and rotation of the polarization of ultrasound.

In [214] and [218] equations (6.85) and (6.86) were used to evaluate the parameters characterizing the electronic subsystem of tungsten by fitting calculated curves to the experimental ones. To accomplish this the following values were used: $s_0 = 2.88 \times 10^5\,\mathrm{cm/s}$ and $c_{1313} = 1.5 \times 10^{12}\,\mathrm{dyn/cm^2}$ from [72], $\Lambda_e = -8 \times 10^{-12}\,\mathrm{erg}$ from [219], $m_1 = 0.54m_0$, $m_2 = 0.29m_0$, $m_3 = 0.22m_0$ ($m_0$ is the free-electron mass), $\mathcal{E}_F = 4.36 \times 10^{-13}\,\mathrm{erg}$, and $v_F = 6 \times 10^7\,\mathrm{cm/s}$ calculated from the data of [156], the $g$-factor for the ellipsoid holes $g_e = 1.73$ from [220]. This led to $\Omega_0/\Omega_e \approx 0.3$, $\zeta \approx 0.02$.

Any deviation of $\mathbf{H}$ from the [001] crystallographic axis causes beats in the oscillations. These beats stem from the appearance of different periods from the ellipsoids at the $N'$ and $N''$ points. To improve the precision of the field orientation a specimen holder was used that made it possible to adjust the configuration of the magnetic field and the [001] axis. Figure 6.71

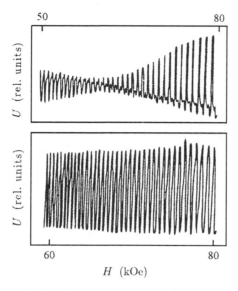

FIGURE 6.71. Amplitude of the signal at the receiving piezoelectric transducer versus the magnetic field in tungsten. (a) Deviation of a few degrees between the magnetic field direction and the [001] crystallographic axis. (b) Parallel configuration to the best accuracy possible in this experiment. The remaining parameters are as in Fig. 6.69. (After Fig. 4 in [214].)

shows that conditions were achieved such that the received signal displayed no beats in the entire range of magnetic field available for the experiment.

The fitting results are shown in Figs. 6.72 and 6.73. The optimum values of the fitting parameters were: $C_1 = 3.6 \times 10^{-6}\,\mathrm{s}^{-1}$ (determined from $\omega_s = C_1 H^3$), $C_2 = 2.7\,\mathrm{s}\,\mathrm{Oe}^2$ (from $\tau_s = C_2/H^2$), $\Delta = 12.7$, and $T = 2.7\,\mathrm{K}$, which differed slightly from the temperature of the experiment since the Dingle temperature, which takes into account the finite relaxation time of the electron subsystem, should be used in equations (6.80) and (6.81). The corresponding relaxation time was estimated to be $\tau \approx 10^{-11}\,\mathrm{s}$. Note that it is characteristic of the specific group of charge carriers: holes on the central cross-section of the ellipsoids. The fitting procedure was carried out only for oscillatory parts of the wave numbers. The monotonic contributions were taken to be those measured in the experiment. Since the form of the observed oscillations was close to harmonic, the terms with $n = 1$ and $n = 2$ in (6.80) and (6.81) were taken into account.

It is seen that the calculations performed for the three field intervals give good agreement with the experimental results. Thus, although experimental investigations of quantum oscillations are very difficult (the necessity of precisely adjusting the configuration of field and crystallographic axis, the large amount of experimental data, the long fitting procedure, etc), they provide important information about the electron subsystem of a metal: an averaged component of the deformation potential, the relaxation time

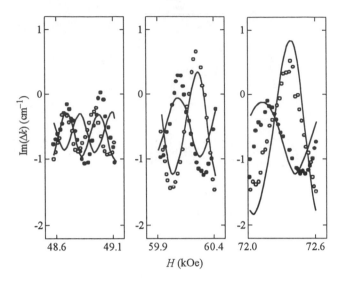

FIGURE 6.72. Magnetic field dependence of the imaginary part of the wave numbers of circularly polarized elastic-like waves in tungsten. Solid lines show the results of calculations performed on the basis of equation (6.86) Filled circles are experimental data for (−) polarization, open ones for (+) polarization. (The remaining paramaters are the same as in Fig. 6.69). (After Fig. 5 in [214].)

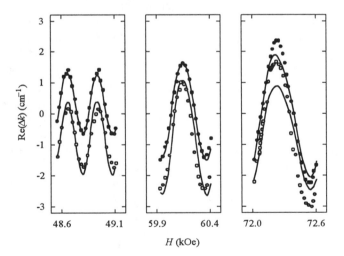

FIGURE 6.73. Magnetic field dependence of the real part of the wave numbers of circularly polarized elastic-like waves in tungsten. The solid lines are the results of calculations performed using equation (6.86). Filled circles are experimental data for (−) polarization, open ones for (+) polarization (the remaining parameters are as in Fig. 6.69). (After [214].)

(specifically for the effective charge carriers), and the shape of the FS, since it shapes the spectrum of the electromagnetic waves.

## 6.7   The Cotton–Mouton Effect in the Classical High-Field Limit

Most of the investigations of polarization phenomena in metals have been carried out with the magnetic field along the direction of propagation and are therefore related to the Faraday effect. In our survey of the published literature we found only one experiment carried out in a non-magnetic metal using the Cotton–Mouton configuration of $k \perp H$. It was performed by Burma et al. [221] in gallium at helium temperatures.

A wave propagating along a principal crystallographic axis is characterized by polarization of the mode along the principal axes. Application of a magnetic field perpendicular to the direction of propagation causes the direction of the polarization plane to be a function of the magnetic field intensity. The experiments are usually performed with a fixed orientation of $H$. However, those using a fixed size but variable direction of the magnetic field, while keeping $H \perp k$, should also be treated as acoustic analogs of the Cotton–Mouton effect.

Burma et al. used a shear wave propagating along the [100] axis of gallium with two different choices for the polarization $u_i$ at $H = 0$: parallel to [001], and parallel to [010]. The receiving transducer was attached so that it responded to vibrations with polarization $u_m$ parallel to [010] in the first case and parallel to [001] in the second. The external magnetic field $H$ was rotated in the (100) plane with the size of $H$ fixed at a value satisfying the classical high-field limit, $kR_H \ll 1$, where $2R_H$ is the diameter of the cyclotron orbit of the charge carriers. The relative value of the received signal amplitude, $A = U_m/U_i$, which corresponds to $\tan \phi$, is shown in Fig. 6.74. These data were used to evaluate the off-diagonal components of the elastic modulus tensor, which proved to be of the order of $10^{-3}$ compared with the diagonal components.

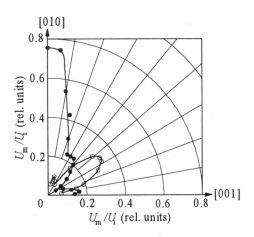

FIGURE 6.74. Polar plot of the relative amplitude of the measured signal for different directions of the magnetic field $\mathbf{H} \perp \mathbf{k} \parallel [100]$ in gallium. Filled circles are for $\mathbf{u}_i \parallel [001]$, $\mathbf{u}_m \parallel [010]$; open circles are for $\mathbf{u}_i \parallel [010]$, $\mathbf{u}_m \parallel [001]$. (After Fig. 1 in [221].)

# References

[1] C. Kittel, Phys. Rev. **110**, 836 (1958).

[2] K. B. Vlasov, Fiz. Met. Metalloved. **7**, 447 (1959) [Phys. Met. Metallogr. (USSR) **7**, No. 3, 121–122 (1959)].

[3] K. B. Vlasov and B. Kh. Ishmukhametov, Zh. Eksp. Teor. Fiz. **36**, 1301 (1959) [Sov. Phys.–JETP **9**, 921–922 (1959)].

[4] K. B. Vlasov and B. Kh. Ishmukhametov, Zh. Eksp. Teor. Fiz. **37**, 745 (1959) [Sov. Phys.–JETP **10**, 531–534 (1960)].

[5] K. B. Vlasov, Fiz. Met. Metalloved. **4**, 542 (1957) [Phys. Met. Metallogr. (USSR) **4**, No. 3, 129–130 (1957)].

[6] K. B. Vlasov, Fiz. Met. Metalloved. **5**, 385 (1957) [Phys. Met. Metallogr. (USSR) **5**, No. 3, 1–6 (1957)].

[7] R. W. Morse and J. D. Gavenda, Phys. Rev. Lett. **2**, 250 (1959).

[8] H. Bömmel and K. Dransfeld, Bull. Am. Phys. Soc. **5**, 357 (1960).

[9] H. Matthews and R. C. LeCraw, Phys. Rev. Lett. **8**, 397 (1962).

[10] B. Lüthi, Phys. Lett. **3**, 285 (1963).

[11] B. Lüthi, Appl. Phys. Lett. **8**, 107 (1966).

[12] Judy R. Franz and Bruno Lüthi, Solid State Commun. **5**, 319 (1967).

[13] R. Guermeur, J. Joffrin, A. Levelut, and J. Penné, Solid State Commun. **5**, 369 (1967).

[14] A. V. Pavlenko, Yu. M. Yakovlev, and V. V. Lemanov, Fiz. Tver. Tela **11**, 3300 (1969) [Sov. Phys.–Solid State **11**, 2673–2675 (1970)].

[15] A. N. Grishmanovskiy, Ph.D. thesis, A. F. Ioffe Physical Technical Institute, St. Petersburg, 1974.

[16] B. K. Jones, Philos. Mag. **9**, 217 (1964).

[17] R. L. Thomas and H. V. Bohm, Phys. Rev. Lett. **16**, 587 (1966).

[18] T. G. Blaney, Philos. Mag. **17**, 405 (1968).

[19] J. D. Gavenda and J. R. Boyd, Phys. Rev. Lett. **15**, 364 (1965).

[20] J. R. Boyd and J. D. Gavenda, Phys. Rev. **152**, 645 (1966).

[21] M. H. Jericho and A. M. Simpson, Philos. Mag. **17**, 267 (1968).

[22] A. R. Mackintosh, Phys. Rev. **131**, 2420 (1963).

[23] Barry I. Miller, Phys. Rev. **151**, 519 (1966).

[24] S. W. Hui and J. A. Rayne, J. Low Temp. Phys. **12**, 49 (1973).

[25] S. W. Hui and J. A. Rayne, J. Phys. Chem. Solids **33**, 611 (1972).

[26] R. Guermeur, J. Joffrin, A. Levelut, and J. Penné, Solid State Commun. **6**, 519 (1968).

[27] C. C. Grimes and S. J. Buchsbaum, Phys. Rev. Lett. **12**, 357 (1964).

[28] L. T. Tsymbal and T. F. Butenko, Solid State Commun. **13**, 633 (1973).

[29] S. V. Medvedev, V. G. Skobov, L. M. Fisher, and V. A. Yudin, Zh. Eksp. Teor. Fiz. **69**, 2267 (1975) [Sov. Phys.–JETP **42**, 1152–1158 (1976)].

[30] A. A. Galkin, L. T. Tsymbal, A. M. Grishin, and T. F. Butenko, Pis'ma Zh. Eksp. Teor. Fiz. **25**, 98 (1977) [JETP Lett. **2**, 87–91 (1977)].

[31] A. A. Galkin, L. T. Tsymbal, T. F. Butenko, A. N. Cherkasov, and A. M. Grishin, Phys. Lett. **67A**, 207 (1978).

[32] K. B. Vlasov and B. N. Filippov, Fiz. Met. Metalloved. **16**, 801 (1963) [Phys. Met. Metallogr. (USSR) **16**, No. 6, 1–7 (1963)].

[33] K. B. Vlasov and B. N. Filippov, Fiz. Met. Metalloved. **17**, 152 (1964) [Phys. Met. Metallogr. (USSR) **17**, No. 1, 142–145 (1964)].

[34] K. B. Vlasov and B. N. Filippov, Fiz. Met. Metalloved. **18**, 333 (1964) [Phys. Met. Metallogr. (USSR) **18**, No. 3, 15–22 (1964)].

[35] V. V. Gudkov and K. B. Vlasov, Fiz. Met. Metalloved. **46**, 254 (1978) [Phys. Met. Metallogr. (USSR) **46**, No. 2, 19–25 (1978)].

[36] K. B. Vlasov and V. V. Gudkov, Fiz. Met. Metalloved. **46**, 892 (1978) [Phys. Met. Metallogr. (USSR) **46**, No. 4, 181–191 (1978)].

[37] K. B. Vlasov and V. V. Gudkov, Pis'ma Zh. Eksp. Teor. Fiz. **28**, 516 (1978) [JETP Lett. **28**, 479–481 (1978)].

[38] A. B. Rinkevich, Metrologiia No. 8, 36, 1982.

[39] V. V. Gudkov, Deposited Article 5349-V90, All-Union Institute for Scientific and Technological Information (unpublished).

[40] V. V. Gudkov and B. V. Tarasov, J. Acoust. Soc. Am. **104**, 2756 (1998).

[41] K. B. Vlasov, Zh. Eksp. Teor. Fiz. **38**, 889 (1960) [Sov. Phys.–JETP **11**, 642–645 (1960)].

[42] V. P. Silin, Zh. Eksp. Teor. Fiz. **38**, 977 (1960) [Sov. Phys.–JETP **11**, 703–707 (1960)].

[43] B. I. Kochelaev, Fiz. Tver. Tela **4**, 1559 (1962) [Sov. Phys.–Solid State **4**, 1145–1147 (1962)].

[44] K. B. Vlasov and B. N. Filippov, Zh. Eksp. Teor. Fiz. **44**, 922 (1963) [Sov. Phys.–JETP **17**, 628–634 (1963)].

[45] K. B. Vlasov and B. Kh. Ishmukhametov, Zh. Eksp. Teor. Fiz. **46**, 201 (1964) [Sov. Phys.–JETP **19**, 142–148 (1964)].

[46] K. B. Vlasov and B. N. Filippov, Zh. Eksp. Teor. Fiz. **46**, 223 (1964) [Sov. Phys.–JETP **19**, 156–161 (1964)].

[47] K. B. Vlasov, Fiz. Met. Metalloved. **19**, 827 (1965) [Phys. Met. Metallogr. (USSR) **19**, No. 6, 26–32 (1965)].

[48] K. B. Vlasov, Fiz. Met. Metalloved. **20**, 3 (1965) [Phys. Met. Metallogr. (USSR) **20**, No. 1, 1–10 (1965)].

[49] K. B. Vlasov and B. N. Filippov, Fiz. Met. Metalloved. **20**, 173 (1965) [Phys. Met. Metallogr. (USSR) **20**, No. 2, 11–15 (1965)].

[50] K. B. Vlasov, Fiz. Met. Metalloved. **21**, 176 (1966) [Phys. Met. Metallogr. (USSR) **21**, No. 2, 17–27 (1966)].

[51] E. R. Muller and J. W. Tucker, Can. J. Phys. **45**, 2443 (1967).

[52] K. B. Vlasov and V. G. Kuleev, Zh. Tekh. Fiz. **37**, 1196 (1967) [Sov. Phys.–Tech. Phys. **12**, 868–873 (1968)].

[53] K. B. Vlasov and V. G. Kuleev, Fiz. Tver. Tela **10**, 2076 (1968) [Sov. Phys.–Solid State **10**, 1627–1634 (1969)].

[54] K. B. Vlasov and V. G. Kuleev, Fiz. Tver. Tela **11**, 877 (1969) [Sov. Phys.–Solid State **11**, 719–726 (1969)].

[55] K. B. Vlasov and V. G. Kuleev, Fiz. Tver. Tela **12**, 1099 (1970) [Sov. Phys.–Solid State **12**, 860–866 (1970)].

[56] K. B. Vlasov and V. G. Kuleyev, Fiz. Met. Metalloved. **32**, 451 (1971) [Phys. Met. Metallogr. (USSR) **32**, No. 3, 1–13 (1971)].

[57] K. B. Vlasov and V. G. Kuleyev, Fiz. Met. Metalloved. **31**, 227 (1971) [Phys. Met. Metallogr. (USSR) **31**, No. 2, 1–13 (1971)].

[58] K. B. Vlasov, V. G. Kuleyev, Ye. V. Rozenfel'd, and M. L. Shur, Fiz. Met. Metalloved. **35**, 5 (1973) [Phys. Met. Metallogr. (USSR) **35**, No. 1, 1–13 (1973)].

[59] K. B. Vlasov and G. A. Babushkin, Fiz. Met. Metalloved. **38**, 936 (1974) [Phys. Met. Metallogr. (USSR) **38**, No. 5, 32–42 (1974)].

[60] G. A. Babushkin and K. B. Vlasov, Fiz. Met. Metalloved. **40**, 695 (1975) [Phys. Met. Metallogr. (USSR) **40**, No. 4, 14–20 (1975)].

[61] K. B. Vlasov and G. A. Babushkin, Fiz. Met. Metalloved. **40**, 939 (1975) [Phys. Met. Metallogr. (USSR) **40**, No. 5, 31–39 (1975)].

[62] K. B. Vlasov and Ya. G. Smorodinskiy, Fiz. Met. Metalloved. **45**, 903 (1978) [Phys. Met. Metallogr. (USSR) **45**, No. 5, 1–7 (1978)].

[63] K. B. Vlasov and Ya. G. Smorodinskiy, Fiz. Met. Metalloved. **46**, 7 (1978) [Phys. Met. Metallogr. (USSR) **46**, No. 1, 1–9 (1978)].

[64] K. B. Vlasov and Ya. G. Smorodinskiy, Fiz. Met. Metalloved. **50**, 1150 (1980) [Phys. Met. Metallogr. (USSR) **50**, No. 6, 21–30 (1980)].

[65] A. I. Akhiezer, V. G. Baryakhtar, and K. B. Vlasov, Usp. Fiz. Nauk **143**, 673 (1984) [Sov. Phys.–Usp. **27**, 641–642 (1984)].

[66] V. V. Gudkov and K. B. Vlasov, Phys. Lett. **103A**, 129 (1984).

[67] V. V. Gudkov and A. V. Tkach, Phil. Mag. Lett. **65**, 267 (1992).

[68] V. V. Gudkov and I. V. Zhevstovskikh, Zh. Eksp. Teor. Fiz. **92**, 208 (1987) [Sov. Phys.–JETP **65**, 117–123 (1987)].

[69] V. V. Gudkov, Solid State Commun. **44**, 229 (1982).

[70] B. W. Roberts, in *Physical Acoustics; Principles and Methods*, edited by W. P. Mason (Academic Press, New York and London, 1968), Vol. IVB, pp. 1–52.

[71] J. Mertsching, Phys. Status Solidi **37**, 465 (1970).

[72] Rohn Truell, Charles Elbaum, and Bruce Chick, *Ultrasonic Methods in Solid State Physics* (Academic Press, New York and London, 1969).

[73] J. W. Tucker and V. W. Rampton, *Microwave Ultrasonics in Solid State Physics* (North-Holland, Amsterdam, 1972).

[74] D. Dominguez, L. Bulaevskii, B. Ivlev, M. Maley, and A. R. Bishop, Phys. Rev. B **53**, 6682 (1996).

[75] L. Onsager, Phys. Rev. **38**, 2265 (1931).

[76] V. D. Fil, P. A. Bezuglyi, E. A. Masalitin, and V. I. Denisenko, Prib. Tekh. Eksp. No. 3, 210 (1973), 1973, [Instrum. Exp. Tech. (USSR) **16**, 905–908 (1973)].

[77] E. A. Masalitin, V. D. Fil, and P. M. Gorborukov, Izmeritel'naia Tekhnika No. 11, 69 (1979).

[78] V. V. Gudkov, Deposited Article 4741-B90 (11.07.1990), All-Union Institute for Scientific and Technological Information (unpublished).

[79] V. V. Gudkov, Deposited Article 1893-B91 (15.01.1991), All-Union Institute for Science and Technological Information, (unpublished), Inventor's Certificate SU 1735760 G01N29/00 (1992).

[80] A. M. Burkhanov, V. V. Gudkov, and I. V. Zhevstovskikh, Prib. Tekh. Eksp. **34**, 167 (1991) [Instrum. Exp. Tech. (USSR) **34**, 658–661 (1991)]. Inventor's Certificate 1516952 (1989).

[81] A. M. Burkhanov, K. B. Vlasov, V. V. Gudkov, and I. V. Zhevstovskikh, Akust. Zh. **34**, 991 (1988) [Sov. Phys.–Acoust. **34**, 569–572 (1988)].

[82] R. N. Thurston, in *Physical Acoustics: Principles and Methods*, edited by W. P. Mason (Academic Press, New York and London, 1964), Vol. 1A.

[83] L. D. Landau and E. M. Lifshitz, *Theoretical Physics: Theory of Elasticity* (Nauka, Moscow, 1987), Vol. 7 (in English: Addison-Wesley, Reading, MA, 1959).

[84] C. Truesdell and R. A. Toupin, in *Handbuch der Physik* (Springer, Berlin, 1960), Vol. III/1, pp. 226–793.

[85] A. I. Akhiezer, J. Phys. USSR **10**, 217 (1946).

[86] M. I. Kaganov and V. M. Tsukernik, Zh. Eksp. Teor. Fiz. **36**, 224 (1959) [Sov. Phys.–JETP **9**, 151–156 (1959)].

[87] C. Kittel and J. H. Van Vleck, Phys. Rev. **118**, 1231 (1960).

[88] L. Landau and E. Lifshitz, Phys. Z. Sowjetunion **8**, 153 (1935).

[89] T. L. Gilbert, Phys. Rev. **100**, 1243 (1955).

[90] N. Bloembergen and S. Wang, Phys. Rev. **93**, 72 (1954).

[91] A. P. Cracknell and K. C. Wong, *The Fermi Surface* (Clarendon Press, Oxford, 1973).

[92] V. M. Kontorovich, Zh. Eksp. Teor. Fiz. **45**, 1638 (1963) [Sov. Phys.– JETP **18**, 1125–1134 (1964)].

[93] V. M. Kontorovich, Zh. Eksp. Teor. Fiz. **59**, 2116 (1970) [Sov. Phys.– JETP **32**, 1146–1152 (1971)].

[94] V. M. Kontorovich, Usp. Fiz. Nauk **142**, 265 (1984) [Sov. Phys.–Usp. **27**, 134–158 (1984)].

[95] V. G. Skobov and É. A. Kaner, Zh. Eksp. Teor. Fiz. **46**, 273 (1964) [Sov. Phys.–JETP **19**, 189–198 (1964)].

[96] H. Stolz and J. Mertsching, Phys. Status Solidi **8**, 847 (1965).

[97] A. I. Akhiezer, Zh. Eksp. Teor. Fiz. **8**, 1330 (1938).

[98] A. V. Tkach, Ph.D. thesis, Institute for Metal Physics, Ekaterinburg, 1994.

[99] P. Aigrain, in *Proceedings of the International Conference on Semiconductor Physics, Prague* (Czechoslovak Academy of Science, Prague, 1961), p. 224.

[100] O. V. Konstantinov and V. M. Perel', Zh. Eksp. Teor. Fiz. **38**, 161 (1960) [Sov. Phys.–JETP **11**, 117–119 (1960)].

[101] R. Bowers, C. Legendy, and F. Rose, Phys. Rev. Lett. **7**, 339 (1961).

[102] D. P. Morgan, Phys. Status Solidi **24**, 9 (1967).

[103] É. A. Kaner and V. G. Skobov, Adv. Phys. **17**, 605 (1968).

[104] B. W. Maxfield, Am. J. Phys. **37**, 241 (1969).

[105] V. T. Petrashov, Rep. Prog. Phys. **47**, 47 (1984).

[106] T. J. Kjeldaas, Jr., Phys. Rev. **113**, 1473 (1959).

[107] J. C. McGroddy, J. L. Stanford, and E. A. Stern, Phys. Rev. **141**, 437 (1966).

[108] A. W. Overhauser and Sergio Rodriguez, Phys. Rev. **141**, 431 (1966).

[109] P. R. Antoniewicz, Phys. Lett. **24A**, 83 (1967).

[110] P. R. Antoniewicz, Phys. Rev. **185**, 863 (1969).

[111] E. I. Kaner and V. G. Skobov, Phys. Lett. **25A**, 105 (1967).

[112] O. V. Konstantinov and V. G. Skobov, Fiz. Tver. Tela **12**, 2768 (1970) [Sov. Phys.–Solid State **12**, 2237–2238 (1971)].

[113] P. R. Antoniewicz, L. T. Wood, and J. D. Gavenda, Phys. Rev. **21**, 998 (1968).

[114] L. T. Wood and J. D. Gavenda, Phys. Rev. B **2**, 1492 (1970).

[115] L. M. Fisher, V. V. Lavrova, V. A. Yudin, O. V. Konstantinov, and V. G. Skobov, Zh. Eksp. Teor. Fiz. **60**, 759 (1971) [Sov. Phys.–JETP **33**, 410–418 (1971)].

[116] V. V. Lavrova, S. V. Medvedev, V. G. Skobov, L. M. Fisher, and V. A. Yudin, Zh. Eksp. Teor. Fiz. **64**, 1839 (1973) [Sov. Phys.–JETP **37**, 929–939 (1973)].

[117] V. P. Naberezhnykh and L. T. Tsymbal, Solid State Commun. **9**, 693 (1971).

[118] V. P. Naberezhnykh, D. E. Zherebchevskii, L. T. Tsymbal, and T. M. Yeryomenko, Solid State Commun. **11**, 1529 (1972).

[119] V. V. Lavrova, V. G. Skobov, L. M. Fisher, A. S. Chernov, and V. A. Yudin, Fiz. Tver. Tela **15**, 3379 (1973) [Sov. Phys.–Solid State **15**, 2245–2251 (1974)].

[120] V. G. Skobov, L. M. Fisher, A. S. Chernov, and V. A. Yudin, Zh. Eksp. Teor. Fiz. **67**, 1218 (1974) [Sov. Phys.–JETP **40**, 605–612 (1975)].

[121] T. F. Butenko, V. T. Vitchinkin, A. A. Galkin, A. M. Grishin, V. A. Mishin, L. T. Tsymbal, and A. N. Cherkasov, Zh. Eksp. Teor. Fiz. **78**, 1811 (1980) [Sov. Phys.–JETP **51**, 909–918 (1980)].

[122] V. V. Lavrova, S. V. Medvedev, V. G. Skobov, L. M. Fisher, A. S. Chernov, and V. A. Yudin, Zh. Eksp. Teor. Fiz. **66**, 700 (1974) [Sov. Phys.–JETP **39**, 338–344 (1974)].

[123] V. G. Skobov, in *Waves and Interactions in Solid State Plasmas*, edited by P. M. Platzman and P. A. Wolff (Mir, Moscow, 1975).

[124] G. Akramov, Fiz. Tver. Tela **5**, 1310 (1963) [Sov. Phys.–Solid State **5**, 955–958 (1963)].

[125] D. N. Langenberg and J. Bok, Phys. Rev. Lett. **11**, 549 (1963).

[126] John J. Quinn and Sergio Rodriguez, Phys. Rev. Lett. **11**, 552 (1963) Erratum: *ibid* **12**, 65 (1964).

[127] John J. Quinn and Sergio Rodriquez, Phys. Rev. **133**, A1589 (1964).

[128] T. G. Blaney, Philos. Mag. **15**, 707 (1967).

[129] K. S. Viswanathan, J. Phys. F. **5**, L107 (1975).

[130] K. S. Viswanathan and Rajam Sekhar, J. Phys. F. **6**, 993 (1976).

[131] Roschen Idiculla and K. S. Viswanathan, Can. J. Phys. **57**, 353 (1979).

[132] I. P. Krylov, Zh. Eksp. Teor. Fiz. **54**, 1738 (1968) [Sov. Phys.–JETP **27**, 934–943 (1968)].

[133] K. B. Vlasov, N. G. Bebenin, and N. S. Yartseva, Fiz. Nizk. Temp. **13**, 579 (1987) [Sov. J. Low Temp. Phys. **13**, 652–655 (1987)].

[134] N. S. Yartseva, N. G. Bebenin, and K. B. Vlasov, Fiz. Met. Metalloved. **66**, 837 (1988) [Phys. Met. Metallogr. (USSR) **66**, No. 5, 1 (1988)].

[135] F. I. Fedorov, *Theory of elastic waves in crystals* (Nauka, Moscow, 1965) [Plenum Press, New York, 1968].

[136] M. J. P. Musgrave, Proc. Roy. Soc. London **A226**, 339 (1954).

[137] M. Ya. Azbel' and E. A. Kaner, Zh. Eksp. Teor. Fiz. **32**, 896 (1956) [Sov. Phys.–JETP **5**, 730–744 (1957)].

[138] V. Ya. Kravchenko, Zh. Eksp. Teor. Fiz. **54**, 1494 (1968) [Sov. Phys.–JETP **27**, 801–808 (1968)].

[139] V. V. Lemanov, A. V. Pavlenko, and A. N. Grishmanovskiĭ, Zh. Eksp. Teor. Fiz. **59**, 712 (1970) [Sov. Phys.–JETP **32**, 389–393 (1971)].

[140] B. V. Tarasov, A. M. Burkhanov, and K. B. Vlasov, Fizika Tverdogo Tela (St. Petersburg) **38**, 2135 (1996) [Sov. Phys.–Solid State **38**, 1176–1180 (1996)].

[141] B. Lüthi, Appl. Phys. Lett. **6**, 234 (1965).

[142] A. N. Grishmanovskii, V. V. Lemanov, G. A. Smolenskii, and P. P. Syrnikov, Fiz. Tver. Tela **14**, 2369 (1972) [Sov. Phys.–Solid State **14**, 2050–2052 (1973)].

[143] K. B. Vlasov, A. M. Burkhanov, and V. V. Gudkov, Zh. Eksp. Teor. Fiz. **91**, 975 (1986) [Sov. Phys.–JETP **64**, 574–578 (1986)].

[144] A. M. Burkhanov, K. B. Vlasov, B. V. Tarasov, and V. V. Gudkov, in *Proceedings of the 1995 Ultrasonic World Congress, Part I* (Gesellschaft für angewandte Ultraschallforschung, Berlin, 1995), pp. 233–236.

[145] A. M. Burkhanov, K. B. Vlasov, V. V. Gudkov, and B. V. Tarasov, to be published.

[146] E. R. Callen, A. E. Clark, B. DeSavage, and W. Colleman, Phys. Rev. **107**, 1735 (1963).

[147] V. L. Gurevich, V. G. Skobov, and Yu. A. Firsov, Zh. Eksp. Teor. Fiz. **40**, 786 (1961) [Sov. Phys.–JETP **13**, 552–555 (1961)].

[148] A. P. Korolyuk and T. A. Prushchak, Zh. Eksp. Teor. Fiz. **41**, 1689 (1961) [Sov. Phys.–JETP **14**, 1201–1202 (1962)].

[149] D. J. Roaf, Philos. Trans. R. Soc. London **A255**, 135 (1962).

[150] E. A. Kaner, V. G. Peschanskii, and I. A. Privorotskii, Zh. Eksp. Teor. Fiz. **40**, 214 (1961) [Sov. Phys.–JETP **13**, 147–155 (1961)].

[151] V. V. Gudkov, Fiz. Met. Metalloved. **66**, 1111 (1988) [Phys. Met. Metallogr. (USSR) **66**, No. 6, 62–69 (1988)].

[152] Roger C. Alig and Sergio Rodriguez, Phys. Rev. **157**, 500 (1967).

[153] A. W. Overhauser, Phys. Rev. **128**, 1437 (1962).

[154] Roger C. Alig, John J. Quinn, and Sergio Rodriguez, Phys. Rev. Lett. **14**, 981 (1966).

[155] Roger C. Alig, John J. Quinn, and Sergio Rodriguez, Phys. Rev. **148**, 632 (1966).

[156] R. F. Girvan, A. V. Gold, and R. A. Phillips, J. Phys. Chem. Solids **29**, 1485 (1968).

[157] K. B. Vlasov, A. B. Rinkevich, and N. A. Zimbovskaya, Fiz. Met. Metalloved. **52**, 517 (1981) [Phys. Met. Metallogr. (USSR) **52**, No. 3, 54–65 (1981)].

[158] I. M. Vitebskii, V. T. Vitchinkin, A. A. Galkin, Yu. A. Ostroukhov, and O. A. Panchenko, Fiz. Nizk. Temp. **1**, 400 (1975) [Sov. J. Low Temp. Phys. **1**, 200–202 (1975)].

[159] L. T. Tsymbal, Yu. D. Samokhin, A. N. Cherkasov, V. T. Vitchinkin, and V. A. Mishin, Fiz. Nizk. Temp. **5**, 461 (1979) [Sov. J. Low Temp. Phys. **5**, 221–225 (1979)].

[160] A. B. Rinkevich, Ya. G. Smorodinskiy, and R. Sh. Nasyrov, Vysoko-chistyie Veschestva No. 3, 42 (1989).

[161] K. B. Vlasov and A. B. Rinkevich, Fiz. Met. Metalloved. **54**, 668 (1982) [Phys. Met. Metallogr. (USSR) **54**, No. 4, 38 (1982)].

[162] V. D. Fil, N. G. Burma, and P. A. Bezuglyi, Pis'ma Zh. Eksp. Teor. Fiz. **23**, 428 (1976) [JETP Lett. **23**, 387–391 (1976)].

[163] N. G. Burma, V. D. Fil, and P. A. Bezuglyi, Pis'ma Zh. Eksp. Teor. Fiz. **28**, 409 (1978) [JETP Lett. **28**, 378–381 (1978)].

[164] J. D. Gavenda and C. M. Casteel, Phys. Rev. Lett. **40**, 1211 (1978).

[165] J. D. Gavenda and C. M. Casteel, Phys. Rev. B **19**, 4331 (1979).

[166] V. V. Gudkov and V. D. Fil, Phys. Status Solidi B **121**, 433 (1984).

[167] H. E. Bömmel, Phys. Rev. **100**, 758 (1955).

[168] R. W. Morse, Bull. Am. Phys. Soc. **1**, 300 (1956).

[169] R. W. Morse and H. V. Bohm, in *Proceedings of the 5th International Conference on Low-Temperature Physics*, edited by J. R. Dillinger (University of Wisconsin, Madison, 1958), pp. 509–512.

[170] R. W. Morse, H. V. Bohm, and J. D. Gavenda, Phys. Rev. Lett. **2**, 250 (1959).

[171] V. V. Gudkov and A. V. Tkach, Philos. Mag. B **68**, 291 (1993).

[172] V. V. Gudkov and A. V. Tkach, Zh. Eksp. Teor. Fiz. **104**, 4109 (1993) [Sov. Phys.–JETP **104**, 985–992 (1993)].

[173] J. A. Rayne and B. S. Chandrasekhar, Phys. Rev. **125**, 1952 (1962).

[174] L. Mackinnon, M. T. Taylor, and M. R. Daniel, Philos. Mag. **7**, 523 (1962).

[175] M. R. Daniel and L. Mackinnon, Philos. Mag. **8**, 537 (1963).

[176] J. J. Quinn, Phys. Rev. Lett. **11**, 316 (1964).

[177] J. J. Quinn, Phys. Rev. **135A**, 181 (1964).

[178] S. G. Eckstein, Phys. Rev. Lett. **12**, 360 (1964).

[179] O. Backman, L. Eriksson, and S. Hörnfeld, Solid State Commun. **2**, 7 (1964).

[180] Y. Eckstein, J. B. Ketterson, and S. G. Eckstein, Phys. Rev. **135A**, 740 (1964).

[181] K. B. Vlasov, A. B. Rinkevich, and A. M. Burkhanov, Fiz. Met. Metalloved. **53**, 295 (1982) [Phys. Met. Metallogr. (USSR) **53**, No. 2, 83–90 (1982)].

[182] A. V. Tkach and A. B. Rinkevich, Fiz. Met. Metalloved. **84**, 79 (1997) [Phys. Met. Metallogr. (USSR) **84**, No. 6, 621–626 (1997)].

[183] G. L. Kotkin, Zh. Eksp. Teor. Fiz. **41**, 281 (1961) [Sov. Phys.–JETP **14**, 201–205 (1962)].

[184] A. H. Nayfeh, *Introduction in Perturbation Techniques* (Wiley, New York, 1981).

[185] B. Casting and H. Goy, J. Phys. C **6**, 2040 (1973).

[186] V. A. Gasparov, Zh. Eksp. Teor. Fiz. **66**, 1492 (1974) [Sov. Phys.–JETP **39**, 732–736 (1974)].

[187] A. B. M. Hoff and J. de Groot, J. Low Temp. Phys. **29**, 467 (1977).

[188] A. V. Golik, A. P. Korolyuk, and V. I. Khizhyi, Solid State Commun. **44**, 173 (1982).

[189] A. V. Golik, A. P. Korolyuk, V. L. Falko, and V. I. Khizhyi, Zh. Eksp. Teor. Fiz. **86**, 616 (1984) [Sov. Phys.–JETP **59**, 359–365 (1984)].

[190] V. V. Gudkov, I. V. Zhevstovskikh, and K. B. Vlasov, J. Phys. F **16**, 739 (1986).

[191] E. Fawcett, R. Griessen, W. Joss, M. J. G. Lee, and J. M. Perz, in *Electrons at the Fermi Surface*, edited by M. Springford (Cambridge University Press, London, 1980), pp. 278–318.

[192] V. A. Gasparov and M. H. Harutuniam, Phys. Status Solidi B **93**, 403 (1979).

[193] R. F. Girvan, A. V. Gold, and R. A. Phillips, J. Phys. Chem. Solids **29**, 1485 (1968).

[194] A. M. Grishin, V. G. Skobov, L. M. Fisher, and A. S. Chernykh, Pis'ma Zh. Eksp. Teor. Fiz. **35**, 370 (1982) [JETP Lett. **35**, 455–458 (1982)].

[195] T. F. Butenko, V. T. Vitchinkin, A. A. Galkin, A. M. Grishin, and L. T. Tsymbal, Zh. Eksp. Teor. Fiz. **78**, 1811 (1980) [Sov. Phys.–JETP **51**, 909–918 (1980)].

[196] V. V. Gudkov and I. V. Zhevstovskikh, Pis'ma Zh. Eksp. Teor. Fiz. **43**, 582 (1982) [JETP Lett. **43**, 752–754 (1982)].

[197] P. K. Larsen and F. C. Greisen, Phys. Status Solidi B **45**, 363 (1971).

[198] V. V. Gudkov, Ph.D. thesis, Institute of Metal Physics of the Russian Academy of Sciences, Sverdlovsk, 1982.

[199] V. V. Gudkov and I. V. Zhevstovskikh, Fiz. Met. Metalloved. No. 9, 34 (1992), [Phys. Met. Metallogr. (USSR) **74**, No. 3, 220 (1992)].

[200] V. V. Gudkov and A. V. Tkach, Phys. Rev. B **52**, 9547 (1995).

[201] V. V. Gudkov and A. V. Tkach, in *Ultrasonic World Congress: Part 1* (Gesellschaft für angewandte Ultraschallforschung, Berlin, 1995), pp. 127–130.

[202] V. V. Gudkov and A. V. Tkach, Czech. J. Phys. **46**, 2537 (1996) Part.S5.

[203] L. Landau, Z. Phys. **64**, 629 (1930).

[204] D. Shoenberg, *Magnetic Oscillations in Metals* (Cambridge University Press, Cambridge, 1984).

[205] L. W. Shubnikov and W. J. de Haas, Proceedings of the Netherlands Royal Academy of Science **33**, 130; 163 (1930).

[206] W. J. de Haas and P. M. van Alphen, Proceedings of the Netherlands Royal Academy of Science **33**, 1106 (1930).

[207] D. H. Reneker, Phys. Rev. Lett. **1**, 442 (1958).

[208] D. H. Reneker, Phys. Rev. **115**, 303 (1959).

[209] J. D. Mavroides, B. Lax, K. J. Button, and Y. Shapira, Phys. Rev. Lett. **9**, 451 (1962).

[210] L. R. Tesrardi and J. H. Condon, in *Physical Acoustics; Principles and Methods*, edited by W. P. Mason (Academic Press, New York and London, 1971), Vol. VIII, pp. 59–94.

[211] Y. Shapira, in *Physical Acoustics; Principles and Methods*, edited by W. P. Mason (Academic Press, New York and London, 1968), Vol. VIII, pp. 1–58.

[212] V. V. Gudkov and I. V. Zhevstovskikh, Fiz. Nizk. Temp. **13**, 976 (1987) [Sov. J. Low Temp. Phys. **13**, 556–557 (1987)]. Erratum: Fiz. Nizk. Temp. **14**, 334 (1988) [[Sov. J. Low Temp. Phys. **14**, 186 (1988)].

[213] M. Lee, J. M. Perz, and D. J. Stanley, Phys. Rev. Lett. **37**, 537 (1976).

[214] V. V. Gudkov, I. V. Zhevstovskikh, N. A. Zimbovskaya, and V. I. Okulov, Zh. Eksp. Teor. Fiz. **100**, 1286 (1991) [Soviet Physics–JETP **73**, 711–716 (1991)].

[215] N. G. Bebenin, N. A. Zimbovskaya, V. I. Okulov, and N. S. Yartseva, Fiz. Nizk. Temp. **15**, 1101 (1989) [Sov. J. Low Temp. Phys. **15**, 613–614 (1989)].

[216] N. A. Zimbovskaya and V. I. Okulov, Fiz. Nizk. Temp. **17**, 728 (1991) [Sov. J. Low Temp. Phys. **17**, 383–386 (1991)].

[217] N. A. Zimbovskaya, V. I. Okulov, N. G. Bebenin, and N. S. Yartseva, Fiz. Met. Metalloved. No. 9, 62-68 (1991)], [Phys. Met. Metallogr. (USSR) **72**, No. 3, 57–63 (1991)].

[218] V. V. Gudkov, V. I. Okulov, I. V. Zhevstovskikh, and N. A. Zimbovskaya, Physica B **211**, 351 (1995).

[219] N. G. Bebenin and N. S. Yartseva, Solid State Commun. **73**, 579 (1990).

[220] J. M. Perz, Can. J. Phys. **55**, 356 (1977).

[221] N. G. Burma, P. A. Bezuglyi, and Ye. D. Radchenko, Fiz. Tverd. Tela **17**, 557 (1975) [Sov. Phys. Solid State **17**, 351–352 (1975)].

# Index